TI SUPPLY

DIVISION OF
TEXAS INSTRUMENTS INCORPORATED

7135 NORTH BARRY AVENUE • ROSEMONT, ILLINOIS 60018

TELEPHONE (312) 296-7187

MOSFET in
Circuit Design

TEXAS INSTRUMENTS ELECTRONICS SERIES

MOSFET in Circuit Design

Metal-Oxide-Semiconductor Field-effect Transistors for Discrete and Integrated-circuit Technology

Robert H. Crawford

Semiconductor Components Division
Semiconductor Research and Development Laboratory
Texas Instruments Incorporated

McGRAW-HILL BOOK COMPANY

New York   San Francisco   Toronto   London   Sydney

1234567890HD72106987

Foreword

Until recently, field effect (unipolar) transistors have had relatively small usage as compared with junction (bipolar) transistors. While the earliest investigations of solid-state amplifying devices were in the direction of field-effect devices (majority carrier devices), the discovery and understanding of minority carrier injection across P-N junctions, led to the subsequent development of bipolar transistors. The bipolar transistor has proved to be an exceptional device, having properties which make it useful over a very broad range of applications. It also effectively made the transition from a two-sided structure to a one-sided structure required for monolithic integrated circuits, and is now the cornerstone of integrated circuits.

Why then, it is natural to ask, is there suddenly a new interest in field-effect devices, and particularly in the metal-oxide-semiconductor (MOSFET) type? The answer is twofold. First is the fact that our technology and understanding of surface oxides such as silicon dioxide (SiO_2) has increased to where very stable MOSFET devices can be reliably produced. Second, and most important, relates to large-scale integrated electronics (LSI). The semiconductor industry is moving rapidly toward integrating much more powerful logic and memory functions on silicon chips than today's production integrated circuits (which are typically three gates in a 14–16 lead package). It is in the domain of LSI that MOSFET technology has generated its greatest interest. The reason is simple— one can put from five to seven times the functional complexity in a given area with MOSFET technology as is possible with bipolar technology. Coupled with this high functional density is the fact that exceptionally high yields can be achieved in the fabrication of large MOS integrated functions because of the relative simplicity of the process.

To go further in the discussion would begin to encroach on the author's excellent book; suffice it to say the area of MOSFET's is one of the most exciting in the semiconductor industry. Robert Crawford has prepared an excellent discussion of the MOSFET device and its application to integrated electronics. The reader will soon recognize that he is being given results of first-hand investigations—Robert Crawford indeed is a key contributor to the Texas Instruments MOSFET program.

This book is another in the Texas Instruments Electronics Series, published by the McGraw-Hill Book Company, in which we are attempting to bring the engineering and scientific audience up to date on subjects of significant current interest.

RICHARD L. PETRITZ, *Director*
Semiconductor Research and Development Laboratory
Texas Instruments Incorporated

Preface

The purpose of this book is to describe basic principles of the MOSFET—particularly as they apply to practical circuit design. The material presented here covers device theory and operation, device characteristics, and finally, device usage in discrete and integrated-circuit form. I have tried to encompass in a single text all the background material required by the design engineer to understand and appreciate MOSFET circuit design.

Material for this book was accumulated over a two-year period during which I was actively engaged in MOSFET circuit design within the MOS program of Texas Instruments Incorporated. It therefore includes much practical information gathered as a result of actual work with MOSFET circuits.

Device coverage within this book has been limited to the MOSFET and does not include the junction FET. There were two reasons for this approach. First, I believe the MOSFET is significant by itself and can stand alone. Second, junction FETS have been covered in an excellent book by L. J. Sevin, Jr.—"Field-effect Transistors," published by the McGraw-Hill Book Company as a part of the Texas Instruments Electronics Series.

The level of presentation is aimed at the practicing circuit-design engineer who has the responsibility of designing an MOSFET circuit. A general background in basic semiconductor theory and circuit design is assumed.

Chapter 1 is an introductory chapter written specifically for the person—such as the salesman, marketing man, or manager—who needs a broad understanding of MOSFET properties without going deeply into the theory. The last half of this chapter considers future trends in such areas as MOSFET technology, devices, and circuits.

Chapter 2 presents basic theory and operation of MOS field effects and derives descriptive equations for device behavior.

Chapter 3 considers the accuracy of the fundamental device equations, develops additional functional relationships, and deals analytically with mobility variation as a function of gate voltage.

Chapter 4 contains a description of the transient response of an MOSFET in a circuit.

Chapter 5 deals exclusively with the MOSFET in integrated-circuit form.

Basic concepts are presented, building blocks used in design are discussed, and an actual MOSFET integrated circuit is described in detail.

Chapter 6 covers the area of MOSFET usage in analog circuits and MOSFET-bipolar combinations.

Like all authors, I am indebted to a number of individuals; without their assistance, this book could never have been written. My thanks are due mainly to Dr. R. M. Warner, Jr., and to Dr. J. R. Biard for their suggestions and help during the writing of the manuscript; to Dr. J. P. Mize for many valuable discussions on mobility; to John R. Miller, technical publications manager at Texas Instruments, for his editoral advice; to Louis Bartning for the design of the decoder shown in the text; to Clifford Arnell for the majority of photographs and measured data; to Mrs. Mary Chubin and Mrs. Eloise Ballard for cheerfully typing the manuscript; and, finally, to the management of Texas Instruments for providing an atmosphere conductive to the writing of this book.

Robert H. Crawford

Contents

Notation

A—area

A_v—voltage gain

β—"gain" term containing device constants; see Eq. (2-15). β_p is positive for P-channel devices, and β_n is negative for N-channel devices.

$\beta(V_G)$—indicates a dependency upon gate voltage

β_o—initial, low-voltage value

β_R—β_D/β_l; see Eqs. (5-9) and (5-10).

C—capacitance

C_o—gate capacitance corresponding to the oxide layer over the channel area; $C_o = A(\epsilon_{\mathrm{ox}}/t_{\mathrm{ox}})$.

E_c—conduction-band energy level

E_F—Fermi energy level

E_i—intrinsic energy level; assumed to be at the center of the gap.

E_{ox}—oxide electric field

E_v—valence-band energy level

ϵ_{ox}—oxide dielectric constant; assumed as $\frac{1}{3}$ pF/cm (relative dielectric constant $\cong 4$).

ϵ_s—silicon dielectric constant; assumed as 1 pF/cm.

g_{ds}—drain conductance in the saturated region

g_{dt}—drain conductance in the triode region

g_m—transconductance $\equiv \dfrac{\partial I_D}{\partial V_G}\bigg|_{V_D}$

g_{mBG}—back-gate transconductance

g_{mD}—transconductance of the driver device

g_{ml}—transconductance of the load MOS

g_{mo}—transconductance at I_{DSS} in a depletion-mode device

IC—integrated circuit

I_D—drain current in the external terminal

I_{DP}—drain current at the point of pinchoff

I_{DSS}—drain current that flows when the gate is returned to the source

J_C—channel current density

k—Boltzmann's constant

K_1—constant; see Eq. (2-24).   (+ for P channel, − for N channel)

L—effective channel length (in direction of current flow)

L'—length of the pinchoff region, measured from the edge of the channel at pinchoff to the edge of drain junction

L_T—channel length, measured from source junction to drain junction

m—voltage parameter in switching-time analysis; see Eq. (4-10)

N_A, N_D, N—doping levels

P—heavy P-type diffusion

ϕ_F—Fermi function; the amount the Fermi level is displaced from the intrinsic level or the center of the gap (as measured in the bulk). Units are in volts.   See Eq. (2-25).

ϕ_s—Surface potential; the amount the intrinsic Fermi level, at the surface, has been bent with respect to the Fermi level. ϕ_s is assumed zero in the flat-band case of Fig. 2-1e.

q—electronic charge $\cong 1.6 \times 10^{-19}$ coulomb

Q_A—charge per unit area in the accumulation region

Q_C—charge per unit area in the channel

Q_D—charge per unit area in the depletion region

Q_G—charge per unit area on the metal gate

Q_I—charge per unit area in the inversion region

Q_{SS}—See approximation No. 9 Chap. 2

r_{ds}—drain resistance, saturation region

r_{dt}—drain resistance, triode region

R_p—parasitic resistance

$R_1\|R_2$—designation for R_1 in parallel with R_2

SR—shift register

θ—a constant in the mobility equation (3-28); $\theta = \beta_o R_P$

τ—time constant

t_{ox}—oxide thickness

T—temperature

μ—mobility; units are in cm²/V-sec. μ_p has a positive sign; μ_n has a negative sign.

$\mu(V_G)$—indicates a dependency upon gate voltage

μ_o—initial low-voltage valve

$v(t)$—voltage as a function of time

V_{BG}—back-gate bias

V_D—voltage at the external drain terminal

V'_D—intrinsic drain voltage, voltage at the internal drain terminal

V_{DD}—drain supply voltage

V_{DS}—drain-to-source voltage

V_G—voltage at the external gate terminal

V_{GG}—gate supply voltage

V_{GS}—gate-to-source voltage

V_{Itk}—intrinsic threshold voltage; defined in Eq. (2-24)

V_P—pinchoff voltage; see Eq. (2-18).

$V_P(v_o)$—pinchoff voltage as a function of output voltage

$V_P(V_{BG})$—pinchoff voltage as a function of back-gate bias

V'_S—intrinsic source voltage, voltage at the internal source terminal

V_{ss}—that portion of the threshold voltage due to Q_{ss}

V_{th}—threshold voltage; see Eqs. (2-22) and (2-27).

V_{thD}—threshold voltage of the driver device

V_{thI}—threshold voltage of the load device

$V_{th}(V_{BG})$—threshold voltage as a function of back-gate bias; see Eq. (2-30).

$V_{th}(v)_{out}$—threshold voltage as a function of the output voltage

$V(y)$—channel voltage as a function of the distance between source and chain

W—channel width (perpendicular to current flow)

x_A—boundary between the accumulation region and the neutral bulk

x_D—boundary between the depletion region and the neutral bulk

x, y, z—coordinates for the MOS structure; see Fig. 2-3.

MOSFET in
Circuit Design

1

An Introduction to the World
of the MOSFET

The metal-oxide semiconductor (MOS)* field-effect transistor (FET) is a voltage-controlled device that exhibits an extremely high input resistance (in the range of 10^{12} to 10^{14} Ω). Unlike the junction FET, the MOS, with its insulated gate, maintains a high input resistance without regard to the magnitude or polarity of the input gate voltage. Even at elevated temperatures, the gate leakage is negligible—thus allowing the use of very large gate bias resistors in analog circuits or direct coupling in digital circuits.

Construction of a P-channel MOS transistor is illustrated in the scale drawings of Fig. 1-1. Two highly doped P-type areas ($10^{18}/cm^3$ to $10^{20}/cm^3$ at the surface) are diffused into an N-type silicon substrate (1 to 10 Ω-cm). These two diffusions are referred to as the *source* and *drain* and are located in close proximity to each other (approximately 0.2 mil separation for a driver device and 1 to 2 mils separation for a load device). A thin (800 to 2,000 Å) insulating material, usually some type of silicon oxide, is placed over the surface of the silicon between the source and the drain, forming the gate dielectric material. Metal is evaporated over the surface of the slice, forming contacts, interconnecting leads, and the gate electrode.

Because of the conditions created by the interfacing, at the surface, of the silicon and oxide, usually all N-channel devices are initially on (at zero gate bias) and all P-channel devices are initially off. Since it is desirable to use an initially off device for switching or digital circuits, at present all commercial MOS inte-

---

* Sometimes referred to as MIS, or metal-insulator semiconductor. Although other MOS structures exist, the MOS field-effect transistor dominates today's technology. In this book, the combination of letters "MOS" will refer to the transistor-type structure discussed in this chapter (see Fig. 1-1).

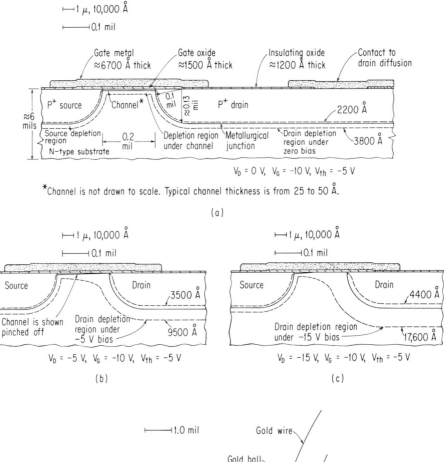

├─┤ 1 μ, 10,000 Å

├──────┤ 0.1 mil

Gate metal ≈6700 Å thick   Gate oxide ≈1500 Å thick   Insulating oxide ≈1200 Å thick   Contact to drain diffusion

P⁺ source   Channel*   0.1 mil   ≈0.13 mil   P⁺ drain   2200 Å

≈6 mils

Source depletion region   0.2 mil   Depletion region under channel   Metallurgical junction   Drain depletion region under zero bias   3800 Å

N-type substrate

$V_D = 0$ V, $V_G = -10$ V, $V_{th} = -5$ V

*Channel is not drawn to scale. Typical channel thickness is from 25 to 50 Å.

(a)

├─┤ 1 μ, 10,000 Å

├──────┤ 0.1 mil

Source   Drain   3500 Å

Channel is shown pinched off   Drain depletion region under −5 V bias   9500 Å

$V_D = -5$ V, $V_G = -10$ V, $V_{th} = -5$ V

(b)

├─┤ 1 μ, 10,000 Å

├──────┤ 0.1 mil

Source   Drain   4400 Å

Drain depletion region under −15 V bias   17,600 Å

$V_D = -15$ V, $V_G = -10$ V, $V_{th} = -5$ V

(c)

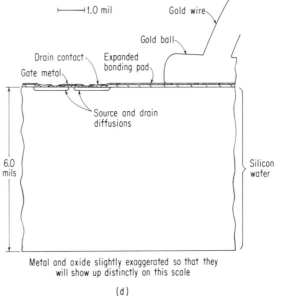

├──────┤ 1.0 mil   Gold wire

Gold ball

Drain contact   Expanded bonding pad

Gate metal

Source and drain diffusions

6.0 mils   Silicon wafer

Metal and oxide slightly exaggerated so that they will show up distinctly on this scale

(d)

Fig. 1-1. Scale drawing of a typical MOS structure in cross section: (a) $V_D = 0$ V; (b) $V_D = -5$ V; (c) $V_D = -15$ V; (d) change in scale showing the MOS in relation to the whole silicon wafer. The channel shown is exaggerated in the depth dimension. The source and substrate are both considered at ground potential.

Fig. 1-2.   MOS symbol.

grated circuits (ICs) are single-polarity P-channel units.   For this reason, this book will deal almost exclusively with P-channel devices.

Figure 1-2 illustrates the symbol for the MOS (P-channel, enhancement-mode) that is used in this book.   The source is the reference terminal, the gate is the control electrode, and the drain is the output of the device.   These three leads are roughly analogous to the bipolar's emitter, base, and collector, respectively.

1-1   OPERATION OF THE MOS

With the drain and source grounded, the gate controls the charge in the *channel* —the region at the substrate surface between the source and the drain.   A negative bias applied to the gate modifies conditions within the silicon.   As the gate accumulates a negative charge, free electrons that are present in the N-type silicon are repelled, forming a depletion region.   Once sufficient depletion has occurred, additional gate bias attracts positive mobile holes to the surface. When enough holes have accumulated in the channel area, the surface of the silicon changes from electron-dominated to hole-dominated material and is said to have *inverted*.   Thus the situation now exists where the two P-diffused regions are connected together by a P-type inversion layer, or channel (hence the nomenclature "P-channel device").   A signal on the gate can modulate the number of carriers within the channel region, so that the gate, in effect, controls current flowing in the channel.   For low values of drain voltage, the inversion layer extends across the entire channel, connecting the drain and source.   Under this bias condition, drain current depends upon drain voltage as well as gate voltage. Notice that in Fig. 1-1a all the P regions, diffused or inverted, are isolated from the substrate material by a depletion layer.

For a constant gate voltage, an increase in the drain voltage alters the situation in the channel region.   Drain current produces an IR drop along the channel. This drop is of such a polarity as to oppose the field within the oxide, produced by the gate bias.   When the IR drop reaches a value to just reduce the field such that an inversion layer is no longer formed, the channel *pinches off* and the drain current tends to saturate at a constant value (independent of drain voltage). The device is said to be *in saturation*.   As can be seen from Fig. 1-1b, the inversion layer is thickest at the source and decreases to zero thickness at the point of pinchoff.*   The voltage across the gate oxide just at the point of saturation is

---

* Actually, the channel cannot go to zero thickness anywhere, for if it did, there would be no current, and there is, in fact, current when the device is in saturation.   "Letting the channel go to zero" is an approximation that allows one to define pinchoff or threshold voltage.

called the *pinchoff* or *threshold* voltage. Threshold voltage can be defined as the voltage across the gate oxide necessary to just produce inversion in the channel.

Further increases in drain voltage drive the MOS harder into saturation. This is demonstrated in Fig. 1-1c, which shows an increase in the depletion regions associated with the drain and a reduction in channel length. Too much of an increase in drain voltage can cause the drain depletion region to *punch through* all the way to the source, resulting in unrestricted current flow if it is not limited by the external circuit.

1-2 CHARACTERISTIC CURVES

When the drain *V-I* characteristics are plotted for a family of gate voltages, the result is similar to the curves in Fig. 1-3. The important features of this figure are:

1. The control parameter is a voltage—as opposed to a current in the bipolar case. This implies high input resistance.
2. The input voltage and the output voltage and current all have the same sign, which allows for the convenient cascading of stages in digital circuits. A junction FET is an example of the opposite case.
3. In Fig. 1-3a, -4 V must be applied to the gate before significant current flows. This characteristic is termed *enhancement-mode operation*. Figure 1-3b illustrates the *depletion-mode* case, where an initial current of -85 μA flows at zero gate bias.
4. The output-current variation for a given increment of gate voltage increases as the gate voltage is increased above the threshold voltage.

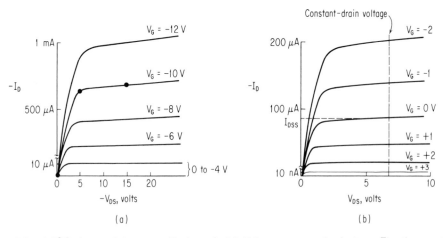

Fig. 1-3. MOS characteristic curves, P channel: (*a*) Enhancement-mode device. The three points on the curves represent the three operating points of Fig. 1-1. (*b*) Depletion-mode device.

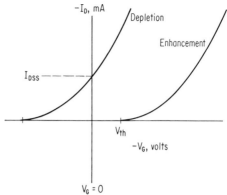

Fig. 1-4. Transfer curves showing I_D vs V_G. Drain voltage is held constant as shown in Fig. 1-3b.

It can be shown that the output current is proportional to the square of the input voltage. In fact, the MOS is often referred to as a *square-law* device.

When a constant-drain-voltage line is graphed on the output characteristics, as in Fig. 1-3b, and the output-current–input-voltage relationship along this line is plotted, the resulting figure is known as a *transfer curve*. The transfer curves of Fig. 1-4 illustrate clearly enhancement-depletion-mode operation, threshold voltage, and the square-law behavior of the MOS. Actual data of drain current plotted as a function of input gate voltage are presented in Fig. 1-5. To illustrate

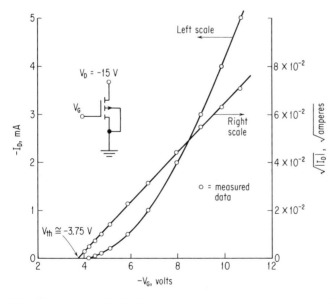

Fig. 1-5. Transfer curve illustrating the square-law characteristic of the MOS.

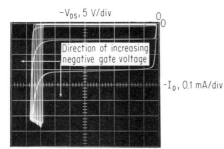

$-V_{DS}$, 5 V/div

0

Direction of increasing
negative gate voltage

$-I_D$, 0.1 mA/div

Fig. 1-6. Characteristic breakdown curves showing the control that the gate voltage exerts over the breakdown voltage.

the parabolic nature of the transfer curve (in saturation), the data in Fig. 1-5 are also presented in the form of $\sqrt{I_D}$ vs. V_G. The result is a straight-line plot. Extrapolating this plot back to where $I_D = 0$ gives the threshold or pinchoff voltage*—the turn-on voltage that is of interest to the circuit designer.

By increasing the magnitude of the drain voltage sufficiently, breakdown characteristics† similar to those in Fig. 1-6 can be seen. Observe that breakdown is a function of gate voltage. The magnitude of the breakdown voltage increases as the gate voltage is made more negative and decreases as the gate voltage is made more positive.

1-3 DIGITAL CIRCUITS

Usually, complex MOS ICs consist only of MOS transistors—no resistors, capacitors, or diodes—as functional elements. The basic building blocks of MOS circuitry are simple NAND and NOR gates such as those shown in Fig. 1-7. These gates, singly and in combination, are used to implement logic-design equa-

---

* It has been suggested that the term "pinchoff" be restricted to depletion-mode devices and the term "threshold" be applied to enhancement devices. Pinchoff voltage is reminiscent of the junction FET (a depletion-mode device) and, as such, has some precedent established for its use. In common parlance, the pinchoff voltage of a depletion device is the gate voltage required to suppress the drain current from its initial value of I_{DSS} to approximately zero. This is shown on the characteristic curves of Fig. 1-3b as approximately +3 V.

With the advent of the P-channel enhancement device, the term "threshold" has become prevalent in field-effect terminology, the threshold being roughly that gate voltage necessary to initiate conduction.

A problem exists in that the origin of the transfer curve is given separate designations, depending upon whether it is to the left or to the right of $V_G = 0$. (See Fig. 1-4.)

Since the same set of equations describes either enhancement-mode or depletion-mode operation (with only a change in sign for the turn-on voltage) and since the two types of devices operate in essentially the same way, the author will use V_P and V_{th} interchangeably. Because most MOS devices operate in the enhancement mode, the term "threshold" will dominate throughout this book. However, when it clarifies matters to use the term "pinchoff," it will certainly be used.

† Assuming that punchthrough does not exist.

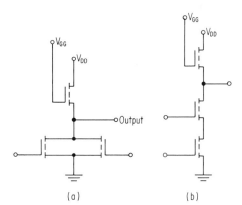

Fig. 1-7. Basic logic gates: (*a*) **NOR,** (*b*) **NAND.**

(a)   (b)

tions, create flip-flops and memory elements, implement timing delays, and in general perform all the functions necessary for digital control.

NOR-logic implementation is the best suited to MOS circuit design. More efficient use of space, as compared with NAND logic, can be cited as the reason. Figure 1-8 shows a combination NAND-NOR gate together with a diagrammatic layout. Because of the series arrangement, Q_1–Q_3 must have three times the gain of Q_4 in order to reduce the total ON resistance to the equivalent of a single

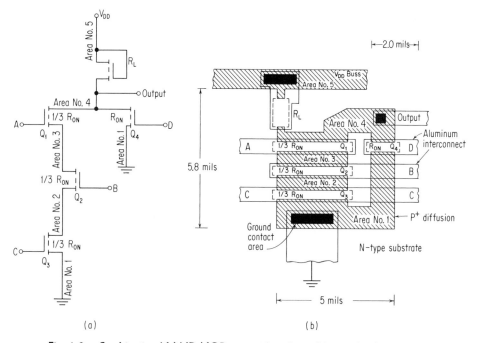

(a)   (b)

Fig. 1-8. Combination NAND-NOR gate with a physical layout (scale drawing).

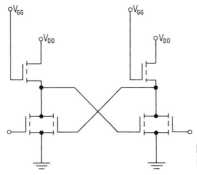

Fig. 1-9. Cross-coupled flip-flop made from two basic NOR gates.

parallel device.   This leads to three times the area per device in the series arrangement shown.

Combining NOR gates, as in Fig. 1-9, results in a cross-coupled flip-flop.   Cross-coupling resistors are eliminated because of the extremely high input resistance. Speedup capacitors are also unnecessary.   Low-gain MOS transistors are used as load resistors.   Values as high as 200 kΩ are obtainable in this manner.   The load gates can be returned either to the drain or to a separate supply.   Reducing V_{DD} and returning the load gates to a separate supply result in lower power and higher-speed operation.

1-4 SPEED

The speed limitation of MOS circuits is due entirely to stray circuit capacitance and the inability of the MOSFET to charge and discharge this capacitance.   Intrinsic cutoff frequencies of MOS devices themselves are in the order of 1 GHz or higher.   Currently, however, one rarely observes commercial MOS circuits in operation much above 2 MHz.   (An exception is four-phase circuitry, which can operate close to 10 MHz.)   In contrast, bipolar ICs often operate an order of magnitude or more faster then MOS circuits.   The basic reason for the difference in speed is that the bipolar has a higher g_m, or gain per unit area, than does the MOS.   A comparison of bipolar-MOS transconductance as a function of device current dramatically illustrates the gain superiority of the bipolar device (see Fig. 1-10).   For typical integrated-device sizes, the bipolar has from 10 to 500 times the g_m of the MOS, depending upon the current level.   The MOS gain can be increased by increasing its width; however, since capacitance as well as g_m is a linear function of area, capacitance also increases.

A typical switching waveform of an MOS inverter is shown in Fig. 1-11.   The turn-on time, controlled by the driver device, is normally much shorter than the turn-off time of the load.   In fact, t_{ON} can generally be ignored in comparison with t_{OFF}.   Two factors contribute to the fact that $t_{OFF} \geq t_{ON}$.   First, the resistance of the load device is typically a factor of 10 greater than that of the driver. This implies that for a given stray capacitance, the time constant for the load is 10

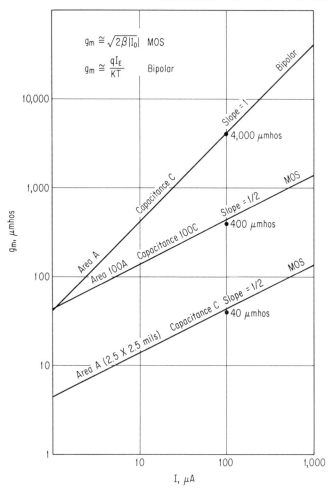

Fig. 1-10.   Comparison of bipolar and MOS transconductance.

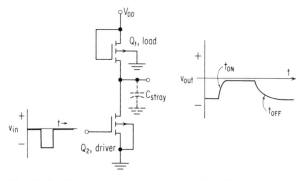

Fig. 1-11.   Basic inverter showing typical switching waveform.

times greater than that for the driver. Second, the gate-to-source controlling bias of the driver remains constant at $-v_{in}$ during switching. In the load case, the control voltage is modulated by the output (source) voltage in such a way as to reduce the gain of the load as the output increases. Q_1 can be described as a nonlinear resistor whose value increases as the voltage across C_{stray} increases. These two factors taken together show that the load charging time is responsible for restricting MOS circuits to low-frequency operation. Fortunately, there are a number of things that can be done to reduce the deleterious transient effects of an MOS load.

1-5   COMPLEMENTARY STRUCTURES

Complementary MOS technology has long been regarded as the ideal solution to many of the difficulties encountered in single-polarity MOS complex ICs. A complementary gate draws approximately zero standby power, making it useful for low-power applications, particularly large, active memories; switching speed is significantly lower than in the case of single polarity; circuit voltages swing the full supply voltage, making two supplies unnecessary; and finally, the output driving impedance is considerably lower than for the single-polarity case. To the circuit designer, these advantages make complementary circuits very attractive. However, complementary technology is not without severe drawbacks.

The major difficulty at the present time is the technology required to fabricate similar complementary devices. N- and P-channel devices inherently have different characteristics when fabricated together; i.e., N-channel devices are depletion mode, while P-channel devices are enhancement mode. The added processing, such as additional diffusions and photomasking steps, necessary to achieve monolithic complementary circuits increases the cost and reduces the yield—undesirable factors in terms of both manufacturer and customer. Additional drawbacks are:

1. Because both polarity devices exist side by side, some form of isolation (not necessary in single-polarity circuits) must be used, resulting in a significant increase in area per function.
2. The number of devices required to implement a given function is greater than in the single-polarity case.

A complementary inverter is shown in Fig. 1-12. Both devices are of the enhancement type. When the input is low, the N-channel device has its gate returned to the source and is off. The gate of the P-channel device, however, is returned to the most negative potential in the circuit (ground) and is thus turned on. Under these conditions, the output goes high and is inverted with respect to the input. When the input goes high, the N-channel device is turned on and the P-channel device is off, resulting in a low output. Notice that in either case, one device is on (presenting a low driving-point impedance to C_{stray}) and the other device is off (limiting the static current drain, and thus the power, to the leakage value).

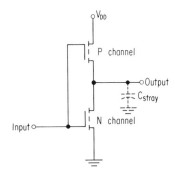

Fig. 1-12. Basic complementary inverter.

1-6 THE FUTURE OF MOS TECHNOLOGY*

For the MOS to have a future, it must be able to offer an advantage over existing bipolar ICs. In general, the advantage will not be in performance. Bipolars, because of their low threshold voltage, high transconductance, and low saturation voltage, will outperform MOS circuits when compared on a speed-power figure-of-merit basis. The advantage offered by MOS technology will be lower cost. This will be achieved through the ability of MOS technology to integrate more functions on a given chip and to give consistently higher processing yields than today's technology allows.

1. Discrete Devices and "Discrete Gates." The demand for discrete MOS devices will be small when compared with that for ICs. Discretes *will* be used in (1) analog switches, because of the theoretically zero offset voltage; (2) high-frequency amplification, where noise is low and the square-law characteristic is desirable; and (3) certain isolated cases, such as the interfacing between MOS-bipolar circuits or where the unique property of the device (i.e., the extremely high input resistance) can be used to advantage.

Along the same line of reasoning, there will be virtually no demand for packaged single-gate functions—the reason being that the performance would be so poor and the cost so close to the cost of present bipolar circuits that no advantage for using MOS would exist.

2. Integrated Circuits. The volume MOS market will be in the area of large, complex ICs operating at slow-to-medium speeds. These ICs will be complete functions representing small systems or subsystems. Integration of a complete function on a chip allows all interconnections to be made on that chip. Thus only the signal and power leads need be brought out of the package. As the number of leads decreases, the package cost decreases. The cost of testing complex ICs, which can make up a very substantial portion of the total cost, is largely dependent upon the number of signal pins in the package. As the number of pins increases, so does the time that is required to test all combinations and permutations. The

---

* All comments in this section apply to single-polarity P-channel ICs, except comment No. 6, which applies to the complementary technology.

ideal situation is the case of a large, complex IC having one input and one output. Here the signal is fed into the circuit, operated upon in some complex manner, modified, and then presented at the output.   The serial shift register (SR) is the classical—and most illustrated—example.   Here a pulse is shifted into the register, stored for a predetermined time (milliseconds, minutes, even days), and then, upon command, presented at the output, ready for use.   Each bit in the SR string drives an identical bit, so that internal wiring and layout problems, capacitance, large output buffers, and operating tests are kept to a minimum.

The SR, although an extremely useful function, represents the extreme case for a minimum number of leads.   More typically, a complex IC has a number of input and output leads.   An example is the binary-to-decimal decoder illustrated in Fig. 5-25.   Here the circuit requires 8 input and 15 output signal leads in addition to the power leads.

3. Custom Design.   As integrated functions on a single chip become larger and more complex, they also become more specialized and therefore more restricted to a given application.   Hence custom-designed circuits will dominate the MOS IC market.   It seems to the author that it is possible for system companies (both large and small) to do their own MOS IC design.   This is definitely *not* the case in bipolar IC design.   One major difference between MOS and bipolar complex IC technology is thus emphasized—the ease with which an idea is reduced to practice is more than an order of magnitude greater in the MOS world.   The customer will do more of the design himself.   A notable exception will be SRs that are sold "by the foot."

The question of who will do MOS design in the future has been given a very colorful answer by Seely and Wanlass in the following quote:[1,*]

> The above considerations have prompted a chemist to remark that he could much more readily design an MOS circuit than he could a conventional transistor circuit to perform the same function.   The same sort of thing holds true when it comes to a comparison of MOS and bipolar integrated-circuit design.   Given the proper design rules, a good digital designer can learn to engineer his own circuits only after a few hours instruction.   This alone almost guarantees the future of MOS, since it puts the burden of design creativity back in the lap of the practicing engineer, where it belongs.   In effect, design engineers will be given a new dimension in which to exercise their creativity.

4. Complex Integrated-circuit vs. Array Technology.†   In general, to gain the greatest advantage from MOS technology, the concept of the *complex IC* will be used. A basic tenet of this concept is the design, layout, and fabrication of the complete function in the smallest possible area.   The function is specialized to do only one job, which it does efficiently.   The unit is handled, tested, and packaged as a

---

* Superscript numbers indicate works listed in the Bibliography at the end of the chapter.
† See Ref. 2 for an expanded discussion.

complete function.   For the function to work, 100 percent of the components within the function must work.   If the function does not work, it is discarded.

On the other side of the fence is *array* technology.   Here the idea is to fabricate a large number of basic building blocks (such as simple gates), test each block, catalog the working units, and interconnect these units (by passing the malfunctioning ones) into complete functional systems or subsystems.   Until now, the main effort of array technology has been directed toward bipolar circuits (with a few notable exceptions).   Now, this does *not* mean that the MOS will not be used in array technology.   It *does* mean that the basic unit cell will be much larger (functionally) than in the present bipolar case.   (Here the unit cell might be a simple four-input gate.)   Consider an example where a customer requires 500 bits of serial SR.   Strings of 50 bits each (300 transistors) might be considered as the basic building block; i.e., compound units of 50 bits are probed as a single unit. Ten working strings would then be connected serially (in a separate masking step) for the required total of 500 bits.

5. Toward Faster Circuits.   Because the speed of the MOS is close to 100 times below its theoretical limit, the next few years will show more dramatic improvements in speed in MOS technology than in bipolar technology.   This is simply a way of saying that it will be easier to achieve an order-of-magnitude increase in speed in the MOS area than in the bipolar area.   Significant speed improvements will come about in the following four areas: (1) circuit innovation, i.e., "tricks" the circuit designer can do to increase performance while not actually changing the process—such as returning the load gates to separate supplies or multiphase clocking schemes; (2) new materials, such as GaAs, which exhibit a higher mobility than silicon; (3) new and improved gate dielectrics that are thinner and have a higher dielectric constant than the currently used silicon oxide; and (4) new technologies, such as dielectric isolation or silicon on sapphire to reduce the stray capacitance or a method to control and increase the surface mobility of currently used materials.

6. Complementary MOS.   Complementary ICs will not have the major impact on the market that single-polarity P-channel circuits have had or will have. Because of the sacrifice necessary in the yield (more processing steps), complementary circuits will be higher-cost items than single-polarity circuits.   Also, with the increased area that is necessary per function, any size or yield advantage over the bipolar is greatly reduced.   Complementary circuitry will thus be forced to compete with bipolars more in the area of performance than in cost.   Complementary MOS technology will make inroads into low-power low-duty-cycle digital functions such as large, active memory applications, where power must be conserved.   One can predict that the complementary MOS will dominate the high-performance, high-cost MOS market, while the single-polarity P-channel MOS will be sought after by the larger, general-purpose, low-cost market.

7. Advances in MOS-device Technology.   So versatile is the MOSFET that numerous new devices, circuits, and techniques will emerge from today's technology.   While the MOS is normally thought of as a low-gain, small-signal device,

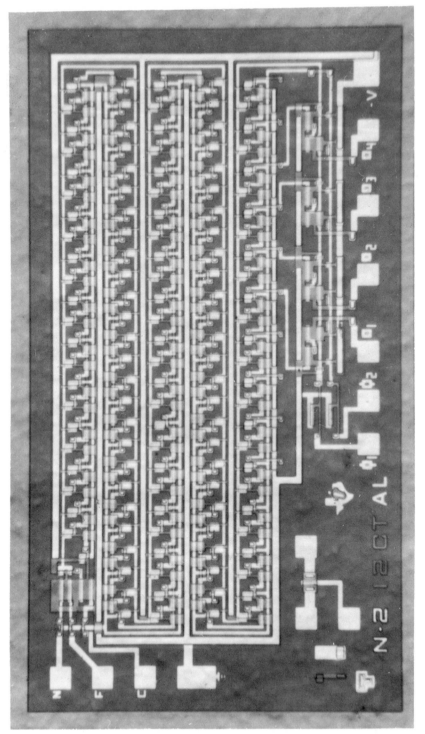

Fig. 1-13.   64-bit SR.

16

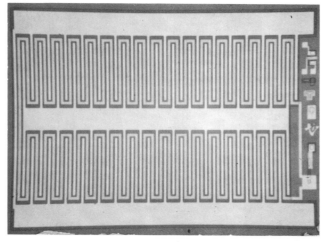

(a)

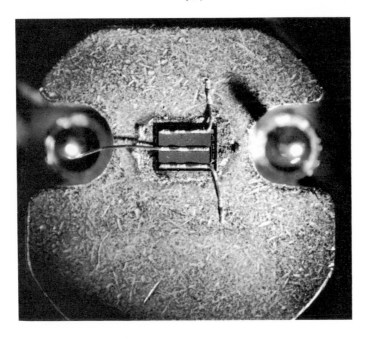

(b)

Fig. 1-14.   MOS power transistor.

practical power devices—capable of 50 to 100 W—have already been demonstrated. Exotic devices such as MOS Thyristors (SCR), transducers, negative-resistance elements, and MOS tunneling devices will begin to appear on the scene.

The wide variety of MOS devices and the many combinations that are possible on a single chip certainly indicate the versatility of this new technology. Linear and digital devices fabricated together will enable functions such as analog-to-digital conversion to be carried out on single monolithic chips. Small signal and power devices will routinely be fabricated together to form complete amplifiers. Combinations of MOS and bipolar technologies will offer improved performance by utilizing the outstanding features of each type of device. Examples of useful combinations are (1) very high MOS resistance values in bipolar circuits, (2) bipolar output buffers in complex MOS ICs, and (3) M-O-S structures in conjunction with a planar bipolar device for improved gain, stability, and breakdown properties of the bipolar device.

8. Examples of Current MOS Device and Circuit Technology. Figures 1-13 and 1-14 illustrate what is currently available in the area of sophisticated MOS circuitry and devices. Figure 1-13 is a photograph of a complex MOS integrated circuit. This unit is a 64-bit dynamic SR requiring two clocks for operation. The SR, consisting of 413 devices, is a single-input serial register with four separate outputs, all on a silicon bar 60 × 100 mils.

A large-geometry MOS power transistor is shown in Fig. 1-14, both in chip form and mounted on a power header. The device exhibits a typical transconductance of 1,000,000 μmhos and requires a chip 90 × 120 mils. Frequency characteristics are excellent as illustrated by the fact that the device switches 1 A < 20 ns. This MOSFET is capable of switching 1 to 3 A at 35 to 45 V.

9. In Summary. MOS complex ICs will make up a significant portion of the IC market within the next few years. The MOS will gain in volume not so much by replacing present bipolar ICs but by expanding the market. Areas will open up that have previously been closed to electronics for economic reasons. Applications, such as digital filters and digital differential analyzers, requiring vast quantities of transistors will now become economically feasible. The MOS will be able to make dramatic inroads into areas that have been dominated by mechanical methods; a prime example is the small desk calculator.

BIBLIOGRAPHY

1. Seely, J. L., and F. M. Wanlass, contributors: *EEE* Specifying Guide: MOS Integrated Circuits, *EEE*, pp. 60–70, May, 1966.
2. Petritz, R. L.: Large Scale Integration Technology, *Trans. Met. Soc. AIME*, vol. 236, pp. 235–249, March, 1966.

GENERAL REFERENCES

Bogert, H. Z.: Metal Oxide Silicon Integrated Circuits, *SCP and Solid State Technol.*, pp. 30–35, March, 1966.

Christiansen, D.: EEE Specifying Guide: MOS Integrated Circuits, *EEE*, pp. 60–70, May, 1966.

Editorial Staff: Planning To Use MOS Arrays? *Electron. Design*, pp. 42–45, Jan. 18, 1966.

Field, R. K.: MOS Arrays Diffuse into Commercial Market, *Electron. Design*, pp. 22–26, Jan. 18, 1966.

Lohman, R. D.: Applications of MOSFET's in Microelectronics, *SCP and Solid State Technol.*, pp. 23–29, March, 1966.

Seely, J. L.: MOS Arrays Have More on a Chip, *Electron. Design*, pp. 90–93, Jan. 4, 1966.

Thornton, C. G.: New Trends in Microelectronics Fabrication Technology 1965–1966, Part I, *SCP and Solid State Technol.*, pp. 42–49, March, 1966.

Warner, R. M.: A Comparison of MOS and Bipolar Integrated Circuits, *NEREM Record*, IEEE Catalog no. F-70, pp. 68–69, November, 1966.

White, M. H., and J. R. Cricchi: Complementary MOS Transistors, *Solid-state Electron.*, vol. 9, pp. 991–1008, October, 1966.

2

Theory of Operation

Because of the nature of this book, a somewhat simplified, first-order-approximation theory will suffice rather than a long and laborious "complete" analysis. A number of approximations will be made which greatly simplify the model of the MOS field effect. This approach will facilitate the understanding of the basic principles of the MOS device. When additional effects, not predicted by the simple theory, are considered, they are discussed separately in order not to make the analysis too large and unwieldy. One can fast become bogged down in excessive algebraic manipulations and lose sight of the real goal of this chapter—an understanding of the workings of the MOS. For those interested in a more comprehensive study of the MOS structure, the author refers the reader to an excellent paper by Ihantola and Moll.[1,*]

The following analysis is divided into two sections: The first gives a qualitative description of the internal workings of the MOS; the second concentrates on a quantitative discussion which, in due course, derives the characteristic equations describing the MOS.

Approximations used to describe the MOS model are as follows:

1. Mobility of current carriers in the channel is constant.
2. The variation of the channel thickness is small along the length of the channel.
3. The thickness of the dielectric over the channel region is assumed to be much greater than the channel thickness.
4. Parasitic resistances (such as in the source) are assumed to be so small as to be negligible.
5. The channel is completely shielded from the drain, so no drain-to-channel feedback exists.
6. Doping of the substrate is uniform and nondegenerate.

---

* Superscript numbers indicate works listed in the Bibliography at the end of the chapter.

7. The drain current consists only of channel current. Leakage currents are neglected.
8. The gate dielectric is considered to be a perfect insulator.
9. Throughout this book, extrinsic conditions which affect the conduction properties—such as oxide traps, silicon surface states, interface energy states, ionic centers within the oxide, and work-function differences—will be lumped together into a single effective charge term, Q_{SS}. Furthermore, Q_{SS} is assumed to be constant and located at the silicon-oxide interface.

2-1 QUALITATIVE ANALYSIS*

Even though the following analysis refers specifically to a P-channel MOS on an N-type substrate (for convenience), the resulting equations are applicable to both N-channel and P-channel devices. There are three distinct conditions or regions occurring within the semiconductor, at the surface, that are important to MOS operation. They are the accumulation, depletion, and inversion regions and are controlled by the external bias on the gate electrode. Generally, for an oxide-passivated surface, surface states or energy states at the silicon-oxide interface act as ionized donors[3] whose effect is the same as a positive applied gate voltage.

Figure 2-1a shows the MOS structure used in this discussion. Drain-to-source voltage is assumed to be so small as to be negligible. The energy-band diagram of a P-channel device under zero applied gate voltage is shown in Fig. 2-1b. Here and in the following band diagrams, the intrinsic level ($P = N = n_i$) is designated as E_i and is assumed to be halfway between the conduction-band energy E_c and the valence-band energy E_v. Because of the positive surface-state charge, negative electrons from within the N-type bulk are attracted to and accumulate at the surface ($x = 0$). Accumulation results in a downward bending of the conduction and valence bands. The closer E_c is bent toward the Fermi level, which is set by the substrate doping, the heavier the surface concentration of electrons becomes. Figure 2-1c illustrates the charge-density distribution. The positive charge per unit surface area (Q_{SS}) must be exactly balanced out by the negative charge accumulated near the silicon surface (Q_A). (Charge distributions are approximated by δ functions.) If a small positive bias is now applied to the gate, additional band bending and accumulation result. Again, the total positive charge must equal the total negative charge so as to maintain charge neutrality ($Q_G + Q_{SS} + Q_A = 0$; see Fig. 2-1d).

If a negative voltage is applied to the gate such that it just counters the effect of Q_{SS}, then no bending of the bands exists, a condition which is known as the *flat-band case* (Fig. 2-1e and f) ($\phi_s = 0$). Further application of a negative gate voltage repels from the channel region the mobile electrons associated with donor centers, causing a depletion region to form. The band diagram and charge picture

---

* This discussion follows, in part, that presented by Grove et al.[2]

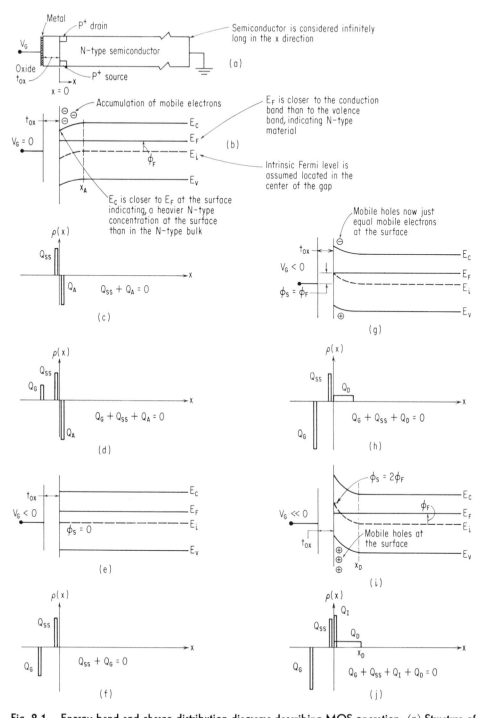

Fig. 2-1. Energy-band and charge-distribution diagrams describing MOS operation: (a) Structure of the device used in this figure. (b) Energy-band diagram for the accumulation condition due to surface states. (c) Charge-density distribution due to surface states. (d) Charge-density distribution due to surface states in addition to positive applied gate voltage. (e) Energy-band diagram for the flat-band case. (f) Charge-density distribution for the flat-band case. (g) Energy-band diagram for the depletion case. (h) Charge-density distribution for the depletion case. (i) Energy-band diagram for the inversion case. (j) Charge-density distribution for the inversion case.

for this case are shown in Fig. 2-1g and h.   When an electron is removed from its donor atom, the ionized atom is left with a net positive charge.   Thus the charge in the depletion region is shown as a positive charge, Q_D, in Fig. 2-1h. When the intrinsic Fermi level E_i is bent just enough to intersect the Fermi level at $x = 0$, the surface has gone from its initial N-type concentration to intrinsic (where $P = N$).   (ϕ_s is now equal to ϕ_F.)   Additional negative gate bias does not extend the depletion region so much as it induces positive mobile holes at the surface.   It is important to note that depletion-region charge and mobile-surface (channel) charge are of the same polarity (positive) and thus their effects add. The sum of these two charge regions must just balance the net charge stored in the oxide (Q_{SS}) and on the gate (Q_G) so as to maintain an electrically neutral system. As the gate bias is increased in the negative direction, a larger and larger percentage of the charge within the semiconductor is contributed by the mobile holes.   (See Fig. 2-1i and j.)

Until $E_i = E_F$, mobile electrons still outnumber mobile holes.   Beyond this point, electrons are suppressed below the intrinsic level while holes in the channel region are raised above this level.   As will be seen later, conduction between the P⁺-type source and drain (by P-type carriers) does not become significant until holes dominate and provide the conducting path.   One can label the onset of conduction (or the apparent onset of conduction, as will be discussed later; see Fig. 2-14) as the threshold voltage V_{th} and define it as where the surface potential goes through intrinsic to a value of $\phi_s = 2\phi_F$.   The threshold voltage is thus the gate voltage that produces a gate charge which just counterbalances the charge contained in surface states and in a depletion region that supports a voltage of $2\phi_F$.   Further increases in gate voltage past V_{th} result in an increase in the number of holes, thus enhancing conduction between the source and drain.

It must be kept in mind that the start of actual conduction is not an abrupt process, in which the channel is completely depleted of carriers at a given gate voltage V_G and then immediately acquires a finite inversion layer at a small increase in the magnitude of V_G.   The minority and majority carriers within the channel are increasing and decreasing continuously and at a finite rate for changes in the gate voltage.   Goldberg described the situation very nicely when he said, "The transition from depletion to inversion is a continuous process, and one must be aware of the fact that the minority carrier concentration increases as the majority carrier concentration decreases and that carriers are always present."[*]

Examination of the gate capacitance as a function of the gate voltage can yield additional insight into the physical operation of the MOS.   Starting with the accumulation condition, Fig. 2-2a shows that the metal gate and the accumulation layer at the silicon surface form a simple parallel-plate capacitor with a separation distance of t_{ox} and a normalized value of $C/C_o = 1$.   Capacitance is measured at the gate, with the source, drain, and substrate grounded.   As negative gate bias is applied, the depletion region that forms tends to separate the two plates

_____

[*] From Ref. 8, p. 594.

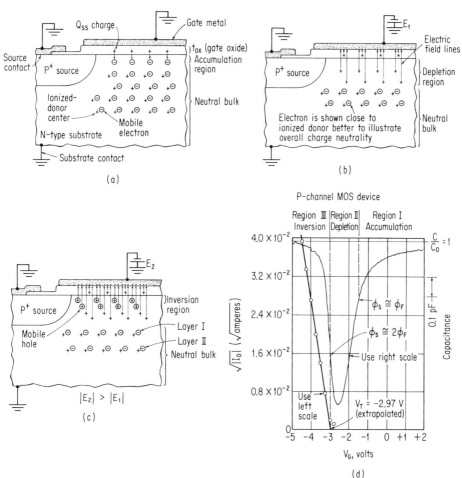

Fig. 2-2. Diagrams illustrating the position of charge under the gate as a function of bias: (a) accumulation case; (b) depletion case; (c) inversion case; (d) C vs. V_G plot of MOS gate showing the three characteristic regions together with a plot of $\sqrt{|I_D|}$ vs. V_G.

of the capacitor and reduce the capacitance (see Fig. 2-2b).   (The two plates are now the gate electrode and the bulk silicon at the edge of the depletion region. Separation is $t_{ox} + x_D$.)   Further application of a negative gate bias attracts holes to the surface (inversion), which in effect decreases the distance separating the two plates, thus increasing capacitance.

A typical C-V plot is shown in Fig. 2-2d.   Region I corresponds to the accumulation region, region II to the depletion region, and region III to the inversion region.   Notice how the capacitance decreases during the transition from accumulation through depletion as the space-charge region forms.   At $\phi_s > 2\phi_F$, the

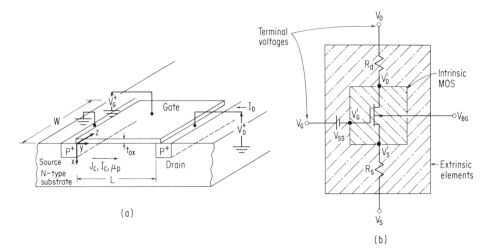

Fig. 2-3. (a) Diagrammatic sketch of an MOS showing dimensions and directions used for analysis. (b) Diagram illustrating the intrinsic and extrinsic elements of an MOS.

charge in the depletion region becomes insignificant compared with the total charge as inversion sets in. The rapid increase in capacitance illustrates how sensitive a function the accumulation layer is to applied gate voltage. Once a significant accumulation layer has formed, the capacitance becomes fixed at approximately $C/C_o = 1$ and is shown to be independent of further gate voltage increases.*

In addition to the C vs. V_G plot, Fig. 2-2d shows the plot of $\sqrt{|I_D|}$ vs. V_G. As will be demonstrated later, the MOS is a square-law device that exhibits a straight-line relationship when the square root of current is plotted as a function of the gate voltage. Extrapolation of this straight-line curve to the point where $I_D = 0$ defines the threshold voltage, or the apparent onset of conduction. In this case, $V_{th} = -2.97$ V. Projecting upward from V_{th} to the capacitance plot shows where the surface potential reaches a value of $2\phi_F$.

2-2 QUANTITATIVE ANALYSIS†

1. Triode Region. Figure 2-3a shows an idealized device together with the coordinate system that is used in the analysis. A brief outline of the "plan of attack" used to derive the device equations is as follows:

1. The channel-current density is integrated over a cross section of the channel ($W\, dx$) to obtain the current.

---

* For a further discussion of C vs. V, see Refs. 2, 5, and 6.
† This discussion follows, in part, that presented by Sah.[3]

2. Channel current is now a function of the charge in the channel.
3. This charge is found by summing all the system charge to zero.
4. Charge is now related to the gate voltage by the use of Gauss' law.
5. The expression for channel current can now be integrated over the length L of the channel.
6. Channel current can now be equated to the external terminal current.

Channel current can be written as

$$I_C = W \int J_C(x,y) \, dx \tag{2-1}$$

where W indicates channel width in the z direction. W is perpendicular to the direction of current flow.

From Ohm's law,

$$J_C(x,y) = \sigma(x)E_y = q\mu_p p(x)E_y \tag{2-2}$$

so that $I_C = W \int q\mu_p E_y p(x) \, dx$, or

$$I_C = Wq\mu_p E_y \int p(x) \, dx \tag{2-3}$$

[μ_p is constant and independent of x (approximation No. 1). μ_p is a positive number, while μ_n is negative.]

Now $E_y = -(dV/dy)$, so that

$$-I_C = W\mu_p \frac{dV}{dy} q \int p(x) \, dx \tag{2-4}$$

where $q \int p(x) \, dx$ represents the mobile charge per unit area in the channel. The problem is now reduced to evaluating $q \int p(x) \, dx$. Since the total MOS-system charge must be zero for a neutral system, the sum of all the charge must equal zero:

$$Q_G + Q_{SS} + Q_C + Q_D = 0 \tag{2-5}$$

where $Q_G + Q_{SS}$ represents all the charge outside of the semiconductor material proper and $Q_C + Q_D$ represents all the charge within the semiconductor material. The channel charge is thus

$$-Q_C = Q_G + Q_{SS} + Q_D \tag{2-6}$$

The charge induced by the gate can be related to the gate voltage by Gauss' law, which states

$$\oint E_{ox} \, dS = \frac{Q_{total}}{\epsilon_{ox}} \tag{2-7}$$

This says that integrating the E field over a given surface (in this case, the gate or channel area) gives the charge under the area, divided by the dielectric constant (gate material in this case). In Eq. (2-7), E_{ox} is considered constant at a given distance y (approximation No. 3) and dS integrates to the differential gate area

$W\,dy$, so that the relation between E and channel charge is

$$\epsilon_{ox}E_{ox}W\,dy = Q_{total} \qquad (2\text{-}8)$$

E_{ox} is defined as

$$-\frac{dV_{ox}}{dx}$$

where dV_{ox} = voltage across the oxide
    dx = oxide thickness

Now
$$-\frac{dV_{ox}}{dx} \cong -\frac{\Delta V_{ox}}{\Delta x}$$

where $\Delta x = +t_{ox}$
    $\Delta V_{ox} = -[V_G - V(y)]$

So that
$$-\frac{dV_{ox}}{dx} = +\frac{V_G - V(y)}{t_{ox}} \qquad (2\text{-}9)$$

(The voltage across the oxide is simply the gate voltage minus the voltage on the channel. Channel voltage will be a function of the distance in the y direction, ranging from V_D at the drain to zero volts at the source.)

Inserting Eq. (2-9) into Eq. (2-8) and letting $\epsilon_{ox}/t_{ox} = C$ (capacitance per unit area) yield

$$Q_G = [V_G - V(y)]C \qquad (2\text{-}10)$$

Equation (2-10) relates the product of the gate capacitance per unit area and the voltage across the oxide to the charge per unit area under the gate.

The gate charge found in Eq. (2-10) is substituted into Eq. (2-6):

$$Q_C = -[V_G - V(y)]C - (Q_{ss} + Q_D) \qquad (2\text{-}11)$$

Equation (2-11) is a mathematical statement of the amount of mobile charge contained in the channel per unit area. Keep in mind that this is the charge that allows conduction between the source and the drain. Q_C will be enhanced by the gate voltage V_G, decreased by the channel voltage created by the drain supply, $V(y)$, decreased by the charge stored in the depletion region beneath the channel, Q_D, and either increased or decreased by Q_{ss}, depending upon its polarity. Later it will be interesting to investigate the condition required to cause the channel charge to go to zero.

Note that $Q_C = q\!\int p(x)\,dx$, which is the desired quantity of Eq. (2-4). Substituting Eq. (2-11) into Eq. (2-4) yields

$$-I_C = W\mu_p\frac{dV}{dy}\{-[V_G - V(y)]C - (Q_{ss} + Q_D)\} \qquad (2\text{-}12)$$

Factoring out C,

$$I_C\, dy = W\mu_p C\, dV \left\{ [V_G - V(y)] + \frac{Q_{ss} + Q_D}{C} \right\} \tag{2-13}$$

Equation (2-13) can now be integrated between 0 and L for the length and between 0 and V_D for the voltage:

$$I_C \int_0^L dy = W\mu_p C \left[V_G \int_0^{V_D} dV - \int_0^{V_D} V(y)\, dV + \frac{Q_{ss} + Q_D}{C} \int_0^{V_D} dV \right]$$

$$I_C L = W\mu_p C \left(V_G V_D - \tfrac{1}{2} V_D{}^2 + \frac{Q_{ss} + Q_D}{C} V_D \right)$$

Now $C = C_o/WL$ and

$$I_C = -\frac{C_o \mu_p}{L^2} [-(V_G - V_{th})V_D + \tfrac{1}{2}V_D{}^2] \tag{2-14}$$

where $V_{th} = -(Q_{ss} + Q_D)/C$. (The threshold-voltage term will be discussed in more detail in Sec. 2-3.)

Equation (2-14) can also be written as

$$I_C = -\beta[-(V_G - V_{th})V_D + \tfrac{1}{2}V_D{}^2] \tag{2-15}$$

where $\boxed{\beta = \dfrac{W \epsilon_{ox} \mu_p}{L t_{ox}}}$ (Because of the sign convention for μ, β_p is positive and β_n is negative.) Now the channel current of Eq. (2-15) is related to the drain current of Fig. 2-3a by the equation $I_C + I_D = 0$ or $I_C = -I_D$. The final form of the equation for the drain current of a device in the triode region can be written as

$$\boxed{I_D = -\beta[(V_G - V_{th})V_D - \tfrac{1}{2}V_D{}^2}\;]^* \tag{2-16}$$

(See Table 2-1 for a summary of MOS-device equations.)

* Equation (2-16) assumes no drain or source resistance (approximation No. 4). At this point, it is easy to include the effects of a given parasitic resistance R_d and R_s. These extrinsic elements are illustrated in Fig. 2-3b, where they are shown to be "outside" of the intrinsic MOS device. Equation (2-13) is integrated from 0 to L on the left-hand side and from V_S' to V_D' on the right-hand side. This integration yields

$$I_D = -\beta[(V_G - V_{th})(V_D' - V_S') - \tfrac{1}{2}(V_D'^2 - V_S'^2)] \tag{2-16a}$$

which represents the triode-region drain current. Equation (2-16a) can be related to the terminal voltages by including the two additional equations

$$V_D' = V_D - I_D R_d \qquad V_S' = I_S R_s$$

The saturation drain current can be written as

$$I_D = -\beta\, \frac{(V_G - V_P)^2}{1 - \beta R_d(V_G - V_P) + \sqrt{1 - 2\beta R_d(V_G - V_P)}} \tag{2-16b}$$

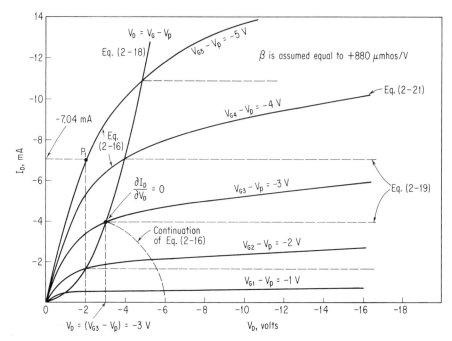

Fig. 2-4. Calculated characteristics of an MOS.

Equation (2-16) was used to generate the triode portion of the characteristic curves for a P-channel device which are shown in Fig. 2-4.   A simple numerical example will illustrate the calculation of point P_1 in the figure.

Assuming $\beta = +880$ μmhos/V, $V_G - V_{th} = -5$ V, $V_D = -2$ V,

$$I_D = -8.8 \times 10^{-4}[+(-5)(-2) - \tfrac{1}{2} \times 2^2]$$
$$I_D = -7.04 \text{ mA}$$

The output current in the triode region depends upon the drain voltage, and hence the drain-source terminals are like a resistor whose value depends upon the gate bias.   Using a small-signal approximation, the resistance is linear and adjustable.   Close to the origin and when $|V_D| \ll |V_G - V_{th}|$, the resistive characteristics are quite linear over a range of voltages and currents.   Since the curves pass through the origin, the characteristics suggest that the MOS can be used as a voltage-controlled d-c, as well as a-c, resistor.   (See Fig. 3-9.)

Equation (2-16) was derived on the basis of the gradual channel approximation (approximation No. 2) and is valid only for the case where $|V_D| \ll |V_G - V_{th}|$.*

---

* A more "exact" analysis by Ihantola[1] points out that, while Eq. (2-16) is valid near the origins, it is only an approximation near the point of saturation.   To improve

When $|V_D| \geq |V_G - V_p|$,† the device enters the saturation or pinchoff region and a different device model applies. Equation (2-16) is also not valid in the low-current region when $V_G \cong V_{th}$. Here the inversion layer is not yet fully formed, and the gate electric field terminates on comparable quantities of mobile charge (in the channel area) and immobile charge (in the depletion region beneath the channel). Because of this fact, the gate voltage no longer controls the channel conductance in the manner stated by Eq. (2-16). The lower boundary on the gate voltage such that Eq. (2-16) is still valid occurs where the channel carrier concentration is just equal to the bulk doping; that is, $p_{\text{channel}} = N_D$. This occurs when the intrinsic Fermi level has been bent such that $\phi_S = 2\phi_F$. (See Figs. 2-1i and 2-2d. Also see Refs. 1, 4, and 7 to 9.) This subject will be covered in more detail in Sec. 2-3.

2. Saturation Region. For a given gate voltage such that $|V_G| > |V_{th}|$, as the drain voltage is increased in magnitude from zero volts, the drain current increases linearly at first, then slows down, and finally tends to level out as $|V_D|$ is made large. The leveling out of the drain current is associated with the pinching off of the channel near the drain. Once the channel has pinched off, the current is said to have saturated at a given level and is then, to a first approximation, independent of drain voltage. Pinchoff occurs because the voltage across the oxide falls below a critical value. The channel IR drop is the factor causing the reduction in electric field. When E_{ox} is decreased to such a value that it cannot support sufficient mobile charge in a given portion of the channel, then that region decreases to (approximately) zero thickness and is said to have *pinched off*. The general shape of the inversion layer is illustrated in Fig. 1-1.

There are two methods for approximating the mathematical boundary between the triode and saturation regions. The first method involves letting the charge in the channel go to zero. This can be seen mathematically by setting Eq. (2-11) equal to zero. Because the channel voltage

$$[V_G - V(y)]C = Q_{SS} + Q_D \tag{2-17}$$

is highest at the drain, pinchoff will begin at the drain as $|V_D|$ is increased. $V(y)$ in Eq. (2-17) can thus be replaced by V_D. Rearranging the terms of Eq. (2-17)

---

the accuracy near saturation or pinchoff, Ihantola gives a more precise equation:

$$I_D = -\beta\{(V_G - V_{th})V_D - \tfrac{1}{2}V_D{}^2 - \tfrac{2}{3}|K_1|[(|V_D| + 2|\phi_F|)^{3/2} - (2|\phi_F|)^{3/2}]\} \tag{2-16c}$$

When $|V_D| \gg 2|\phi_F|$, Eq. (2-16c) reduces to a simpler form:

$$I_D = -\beta[(V_G - V_{th})V_D - \tfrac{1}{2}V_D{}^2 + \tfrac{2}{3}K_1V_D\sqrt{|V_D|}] \tag{2-16d}$$

† See Eq. (2-18).

gives the relationship

$$\boxed{V_D = V_G - V_P}$$ (2-18)

where $V_P = -(Q_{ss} + Q_D)/C.$*

Equation (2-18) states mathematically the boundary line between the triode and saturation regions which is shown plotted in Fig. 2-4. To the right of this line, the MOS is in saturation and $|V_D| > |V_G - V_P|$, while to the left of this line, operation is in the triode region, where $|V_D| < |V_G - V_P|$.

The second method for defining the saturation-triode boundary can be found by examining Eq. (2-16). For a given gate and threshold voltage, the magnitude of the drain current increases as $|V_D|$ is increased from zero volts. Initially, the $(V_G - V_{th})V_D$ term dominates the expression, resulting in the increase of $|I_D|$. Soon, however, the quadratic term becomes significant, with the result that the rate of increase of the current falls off. At some drain voltage, $|I_D|$ will reach a maximum. Past this point, the equation predicts a decrease. However, at the maximum-current point, the device has reached saturation and the device model and equations change. This is why a decrease in current is not seen. To illustrate this point, Eq. (2-16) is plotted past the point $V_D = V_G - V_P$ on Fig. 2-4 for the condition $V_G - V_P = -3$ V. To find the maximum of $|I_D|$ from Eq. (2-16), the expression can be differentiated with respect to the drain voltage and set equal to zero. Solving the resulting expression yields the result already stated in Eq. (2-18), namely, $V_D = V_G - V_P$.

Once saturation has been reached, the voltage drop across the inverted portion of the channel tends to remain fixed at $V_G - V_P$, while V_D varies. To a first approximation, this constant voltage across a constant channel resistance results in a constant drain current. Once saturation has been reached, the output characteristic curves (such as those of Fig. 2-4) can be approximated by horizontal lines. The equation for the current in saturation can be found by placing $V_D = V_G - V_P$ into Eq. (2-16), which results in

$$\boxed{I_D = -\frac{\beta}{2}(V_G - V_P)^2}^{\dagger}$$ (2-19)

* Notice that the pinchoff voltage given is exactly the same as the threshold voltage of Eq. (2-14). Refer to the footnote in Sec. 1-2 regarding pinchoff and threshold voltages.

† Equation (2-19) was derived from Eq. (2-16), and as such it does not include any dependence upon the $\frac{2}{3}|K_1|[(|V_D| + 2|\phi_F|)^{3/2} - (2|\phi_F|)^{3/2}]$ term in Eq. (2-16c). An analysis carried out by Greene and Soldano[10] includes the above term and leads to increased accuracy of Eq. (2-19):

$$I_D = -\frac{\beta}{2}(V_G - V_P')^2$$ (2-19a)

where $V_P' = |V_P| + \frac{2}{3}|K_1|\sqrt{|V_G - V_P|}$.

Equation (2-19) is valid for $|V_D| \geq |V_G - V_P|$.   (See Table 2-1 for a summary of MOS-device equations.)

Equation (2-19) shows a square-law dependence of the drain current upon the gate voltage, and it also indicates that the drain current is independent of the drain voltage.   (See Fig. 2-5 for the square-law transfer characteristic.)   The saturation region, as shown in Fig. 2-4 by dotted lines, is derived from Eq. (2-19). Actual output characteristics of a P-channel MOS device indicate that the drain current is not constant and is, in fact, a function of the drain voltage.   Thus a device model must be developed which will take into account the finite output impedance displayed in Fig. 2-6.

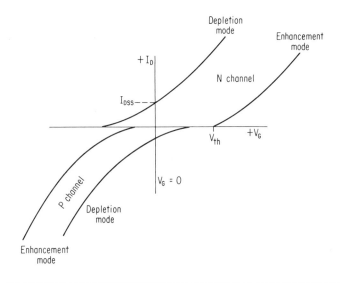

	N channel		P channel	
	Enhancement	Depletion	Enhancement	Depletion
V_{th} or V_p	+	−	−	+
Q_{ss}*	+		+	
Q_D	−		+	
Substrate	P–doped		N–doped	
V_{Ith}	+		−	
V_{ss}	−		−	

*Q_{ss} is generally an ionized donor and thus acts as a positive charge center. The above table was formed on the basis that there were no "added" effects such as channel doping to modify V_{th} and that $|Q_{ss}| > |Q_D|$

Figure 2-5

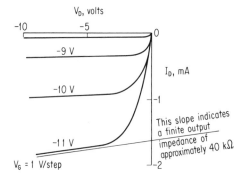

Fig. 2-6. MOS *V-I* characteristics showing finite output impedance.

In an MOS transistor operating in current saturation, there are two major feedback components contributing to charge control in the channel . . . region. One is a direct electrostatic coupling between the drain electrode and the [channel] . . . by way of a near insulating substrate. . . . The other is a modulation of the length of the [channel] . . . with drain voltage.

These two effects are in "parallel"; which effect dominates will depend upon the particular structure under consideration.*

Channel-length Modulation. Figure 2-7 shows a drawing of the channel region used in developing a model for the finite output impedance exhibited by the MOSFET in saturation. Two principal features of the drawing are (1) the depletion width extends into the region of the channel and is a function of the drain voltage, and (2) the channel extending from the source has a voltage developed across it that is independent of the drain potential (to a first approximation). L_T defines the total channel length, from source to drain. L' is the distance from

* From Ref. 11, p. 136.

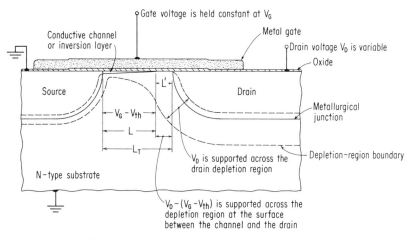

Fig. 2-7. Pictorial model used in the analysis for finite output impedance.

the edge of the channel, at pinchoff, to the drain. $L_T - L'$ represents the effective channel length.

It was stated earlier that the voltage across the pinched-off channel tends to remain fixed at $V_G - V_{th}$, as shown in Fig. 2-7. Any difference in the drain potential and the voltage across the channel must be taken up across a depletion region at the surface, which is designated L'. The voltage supported by this region is $V_D - (V_G - V_{th})$. As V_D increases, L' must also increase to absorb the additional voltage. Thus a modulation of L' by the drain voltage results in a modulation of the effective channel length[1] ($L = L_T - L'$). Increasing the drain voltage decreases the channel length and therefore its resistance. To keep a constant voltage of $V_G - V_{th}$ across the channel, the drain current must increase to compensate for the decrease in channel resistance. This increase in current with an increase in output voltage is positive feedback which results in the finite output impedance illustrated in Fig. 2-6.

Equation (2-19) can be modified to account for the device's finite output impedance. The distance L representing the effective channel length is now replaced by $L_T - L'$, where L' is derived from simple P-N junction theory.

$$L' = \sqrt{\frac{2\epsilon_s[V_D - (V_G - V_{th})]}{qN}} \qquad (2\text{-}20)$$

where N is the impurity density of the substrate. Equation (2-19) is rewritten as

$$I_D = I_{DP}\frac{L_T}{L_T - L'} \qquad (2\text{-}21)$$

where $I_{DP} = -(\beta/2)(V_G - V_{th})^2 \Big|_{V_D = V_G - V_{th}}$ (drain current at the point of saturation or pinchoff).

Equations (2-20) and (2-21) show that as the drain voltage is increased, L' increases, causing an increase in the drain current. A set of output characteristics derived from Eqs. (2-16) and (2-21) is shown in Fig. 2-4.

Hofstein and Warfield[11] have shown that for the case of channel-length modulation, the output conductance is directly proportional to the drain current. This conclusion is verified in Fig. 2-8, where g_{ds} is plotted as a function of I_D. This graph of measured data shows that the drain conductance is, in fact, a linear function of the drain current. Thus one can conclude that for the two devices shown, the dominant feedback mechanism is the channel-length modulation.

Both Hofstein and Warfield[11] and Reddi and Sah[12] conclude that this type of feedback is analogous to the Early effect that one observes in bipolar devices.*

Drain-to-channel Electrostatic Feedback. When the substrate of an MOS device is sufficiently low-doped, the drain field may penetrate into the channel. A model of this situation is shown in Fig. 2-9. Since the charges in the channel cannot tell

---

* The Early effect[13] in bipolar transistors is a modulation of the base width with collector voltage which produces a finite output impedance.

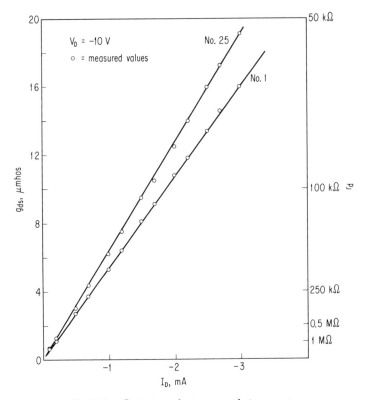

Fig. 2-8. Output conductance vs. drain current.

whether the electric field lines terminating upon them come from the negative-biased drain or the negative-biased gate (P-channel device), a modulation of the drain voltage results in a modulation of the channel conductance.   In addition to collecting carriers coming from the channel, the drain acts as a gate or a control electrode.

When electrostatic feedback dominates, Hofstein and Heiman[14] have shown that

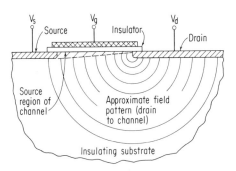

Fig. 2-9. Model illustrating electrostatic feedback from drain diffusion into the channel. (*After Hofstein and Warfield.*[11])

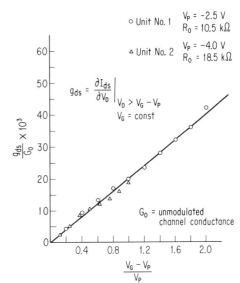

Fig. 2-10. Plot of g_{ds}/G_o vs. $(V_G - V_P)/V_P$. *(After Hofstein and Heiman.[14])*

the drain conductance is directly proportional to the effective gate voltage $(V_G - V_P)$.   Data illustrating this fact are shown in Fig. 2-10.

2-3 ADDITIONAL CONSIDERATION OF THE THRESHOLD VOLTAGE

1. Physical Considerations.   Threshold voltage was given in Eq. (2-14) as

$$V_{th} = - \frac{Q_{SS} + Q_D}{C} \tag{2-22}$$

where Q_{SS} = effective surface-state charge density per unit area

Q_D = bulk charge per unit area associated with the channel depletion region

C = capacitance of the gate to channel per unit area, ϵ_{ox}/t_{ox}

The physical interpretation of Eq. (2-22) is that V_{th} is the gate voltage required to neutralize, in effect, the immobile charge above and below the channel region. Any additional gate voltage over and above $|V_{th}|$ will produce a gate charge that must be neutralized by an equal amount of mobile channel charge.   The charge above the channel is Q_{SS}, while the charge below the channel consists of that within the depletion region located in the bulk (Q_D).

Depletion-region charge Q_D is derived from simple P-N junction theory,* where $N_{\text{channel}} \gg N_{\text{substrate}}$:

$$Q_D = \text{depletion-region thickness } (x_D) \cdot \text{substrate charge density}$$

$$= \sqrt{\frac{2\epsilon_s \phi_s}{qN}} \, qN \qquad \text{where } \phi_s = 2\phi_F \tag{2-23}$$

$$Q_D = \sqrt{2q\epsilon_s N} \, \sqrt{2\phi_F}$$

* See Ref. 15, p. 21.

where $N = N_D$ for an N-type substrate device (P channel)

$N = N_A$ for a P-type substrate device (N channel)

(See the Notation section in the front of the book for the meaning of the other terms.)

The voltage on the gate necessary to support this charge [from Eq. (2-22)] is

$$V_{Ith} \equiv + \frac{Q_D}{C} = -K_1 \sqrt{2\phi_F}^* \qquad (2\text{-}24)$$

where $K_1 = \pm (t_{ox}/\epsilon_{ox}) \sqrt{2q\epsilon_s N}$ (+ for P channel, − for N channel).

Notice that the constant relating gate voltage to charge in the semiconductor is the gate-channel capacitance C ($C = \epsilon_{ox}/t_{ox}$). Equation (2-24) might be considered an "intrinsic" threshold voltage. For a "perfect" device, with zero surface-state charge density ($Q_{SS} = 0$), the MOS device (both N- and P-channel) would be of the enhancement type with a low threshold voltage (in the order of $|0.75|$ to $|1.5|$V) equal to V_{Ith}. (See Fig. 2-3b.)

The term ϕ_F appearing in Eq. (2-23) is the Fermi potential and represents the position of the Fermi level with respect to the intrinsic Fermi level (see Fig. 2-1g). ϕ_F is known as the *contact, diffusion,* or *built-in potential* and is analogous to the same term in a simple P-N junction. From P-N junction theory, ϕ_F is determined by doping level and is given as

$$\phi_F = \frac{kT}{q} \ln \frac{N}{n_i} \dagger \qquad (2\text{-}25)$$

Plots of both Eq. (2-24) and Eq. (2-25) are shown in Fig. 2-11. For typical values where $N_D = 10^{15}$ and $t_{ox} = 1,500$ Å, one can see that the Fermi potential is 0.29 V and the intrinsic threshold is -0.6 V. [Q_D is positive for N-type substrates, and thus Eq. (2-24) yields a negative voltage for a P-channel device.]

Surface-state charge Q_{SS} is assumed (approximation No. 9) to lie close to the silicon-oxide interface[16] and is thus related to the gate voltage through the same constant, C, as is the bulk charge:

$$V_{SS} \equiv - \frac{Q_{SS}}{C} = - \frac{t_{ox}}{\epsilon_{ox}} Q_{SS} \qquad (2\text{-}26)$$

---

* For the simple model presented here of the intrinsic threshold voltage, it is assumed that the depletion region under the channel is depleted to such an extent as to support a constant voltage of $2\phi_F$ across it. (See Ihantola.[1]) At this point ($2\phi_F$), the density of carriers in the channel (holes) is equal to the substrate doping level N_D. (See Terman[4] or Brown.[7]) In this simplified analysis, it is also assumed that the depletion region is uniformly rectangular. To consider more exactly the shape of this region, see Garrett and Brattain.[9] For "exact" calculations, see Goldberg.[8]

A more "exact" calculation of threshold voltage is given by Sah and Pao.[18] In their analysis, $V_{Ith} = |K_1 \sqrt{2\phi_F}| + |2\phi_F|$, which is exactly Eq. (2-24) plus the term $2\phi_F$. A plot of threshold as a sum of these two terms ($K_1 \sqrt{2\phi_F}$ from Fig. 2-11a and $2\phi_F$ from Fig. 2-11b) is shown in Fig. 2-11c.

† See Ref. 15, p. 214.

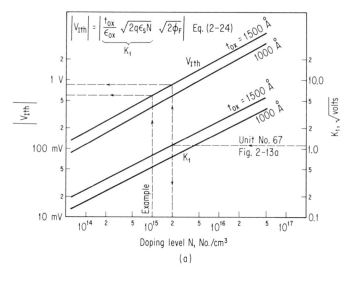

$$\left| V_{Ith} \right| = \left| \frac{t_{ox}}{\epsilon_{ox}} \underbrace{\sqrt{2q\epsilon_s N}}_{K_1} \sqrt{2\phi_F} \right| \quad \text{Eq. (2-24)}$$

(a)

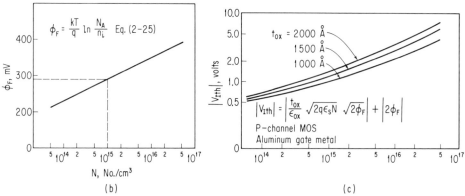

(b)   (c)

Fig. 2-11. (a) V_{Ith}, K_1 vs. N. (b) ϕ_F vs. N. (c) V_{Ith} vs. N. (*After Sah and Pao.*[18])

At the present time, surface-state densities (Q_{SS}/q) in the order of $4 \times 10^{11}/\text{cm}^2$ can be expected.[12]   Because Q_{SS} generally acts as an ionized donor state, Q_{SS} is a positive quantity, giving V_{SS} a negative sign in Eq. (2-26).   (For a summary of the polarities of Q_D and Q_{SS}, refer to Fig. 2-5.)   Using typical values for Eq. (2-26) gives the following threshold voltage due to surface states:*   Assume $t_{ox} = 1,500$ Å and $\epsilon_{ox} \cong \frac{1}{3}$ pF/cm.   Then $V_{SS} = -2.88$ V.

The total threshold voltage is the algebraic sum of the intrinsic threshold voltage and that due to surface states.   From Eqs. (2-24) and (2-26),

$$\boxed{\begin{aligned} V_{th} &= V_{Ith} + V_{SS} \\ V_{th} &= -K_1 \sqrt{\phi_s} + V_{SS} \end{aligned}} \tag{2-27}$$

* To convert the various units of measurement used in MOS work, refer to the conversion table in the Appendix.

where $K_1 = \pm (t_{ox}/\epsilon_{ox}) \sqrt{2q\epsilon_S N}$

$\phi_s = 2\phi_F$

Adding the two voltages found in the previous paragraphs gives a typical threshold voltage for a P-channel device:

$$V_{th} = -0.6 - 2.88 = -3.48 \text{ volts}$$

A P-channel MOS with a negative threshold voltage indicates an enhancement-mode device, as shown in Fig. 2-5. The same device made as an N-channel MOS would turn out as depletion mode:

$$V_{th} = +0.6 - 2.88 = -2.28 \text{ volts}$$

2. Substrate Bias. When a bias is applied to the substrate, modulation of the channel conductance results. Thus the substrate can act as a second gate and is, in fact, sometimes referred to as the *back gate*. Because a P-N junction is formed with the source and drain diffusions to the substrate, some of its characteristics tend to be more like the gate of a junction FET than the gate of an insulated device. The input impedance of this control electrode is much lower than that of the front gate—being on the order of a larger-area reverse-biased diode. Leakage currents, which are almost nonexistent in the front gate, can become quite large, reaching even the microampere range at elevated temperatures. Increasing the magnitude of the reverse bias applied to the substrate will tend to lower $|I_D|$ for a given value of front-gate voltage. Figure 2-12 illustrates this point by showing the output characteristics of a P-channel device as modified by three separate back-gate bias levels.

The theory of operation of the back-gate bias can be explained by an extension of the insulating depletion layer into the substrate, as shown in Fig. 2-2c. As the reverse bias on the substrate increases, the depletion region extends further into the substrate as mobile electrons are depleted and immobile donor centers are uncovered—just as in P-N junction theory. A portion of the electric field lines shown originating from the mobile channel charge (that which contributes to conduction) will be diverted to the "new" uncovered fixed charge. For a fixed gate voltage, there will now be less mobile channel charge and therefore a decrease in conduction. In the model shown in Fig. 2-2c, there are eight mobile holes available for conduction. If the back-gate bias is increased such that layer I becomes uncovered, five fixed positive charges will be uncovered. To maintain a neutral charge system, five channel carriers will disappear, leaving only three for conduction. By applying additional bias to uncover layer II, no more mobile channel carriers will exist, and the current will go to zero.

The expression for the threshold voltage as a function of the back-gate bias can be written directly from Eq. (2-27). One assumption generally used in P-N junction theory is that all the applied voltage will appear across the depletion region and none across the bulk semiconductor. A polarity of back-gate bias V_{BG} is assumed such that when the substrate is reverse-biased with respect to the

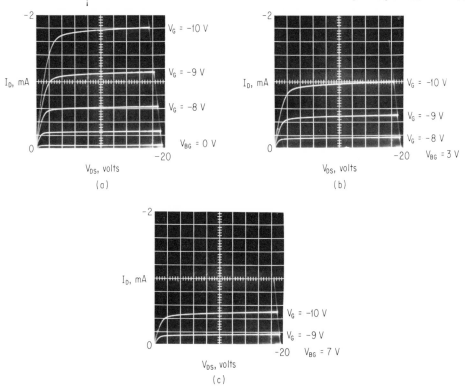

Fig. 2-12.   Output characteristics modulated by a back-gate bias.

source, V_{BG} and ϕ_s have the same polarity.   Modifying Eq. (2-27) to take into account the back-gate reverse bias and using the above assumption give the following result:*

$$V_{th} = -K_1 \sqrt{\pm(\phi_s + V_{BG})} + V_{SS} \qquad (2\text{-}28)$$

where $\phi_s = 2\phi_F$.   (Use of the $\pm$ sign is necessitated by the fact that the term under the radical must be positive.)

* This equation is based upon a simple analysis assuming a one-dimensional field; i.e., the depletion region associated with the channel is thin compared with the length of the channel.   If, however, the depletion-region depth approaches or exceeds the channel length, then the analysis immediately becomes a messy multidimensional problem that the author is not equipped to handle.   Threshold-voltage variations in a few small-geometry devices have been seen to follow a one-third power law, as opposed to the usual one-half power law.   In ICs, the load (and not the driver) is the device affected by back-gate bias modulation.   (See Sec. 5-1.)   As a practical matter, the length of a load device is always large compared with the depletion-region depth (*typical* load lengths are close to 1.0 mil, while depletion-region depths are less than 0.1 mil).   Load devices should be expected to conform to Eq. (2-28) fairly closely.

Equation (2-28) can be verified by plotting the threshold voltage as a function of $-\sqrt{\phi_s + V_{BG}}$, which should result in a straight line. Data of $-V_{th}$ vs. $-\sqrt{\phi_s + V_{BG}}$ are plotted in Fig. 2-13a, which does, in fact, give a straight-line relationship with from 0 to 20 V applied bias. An additional check on the validity of Eq. (2-28) is to plot the log of this equation:

$$\log(-)(V_{th} - V_{SS}) = \tfrac{1}{2}\log(2\phi_F + V_{BG}) + \log K_1 \qquad (2\text{-}29)$$

The "$\tfrac{1}{2}$" term comes from the square root in Eq. (2-28). If the half-power term is correct, then $\log(-)(V_{th} - V_{SS})$ vs. $\log(2\phi_F + V_{BG})$ should show a slope of $\tfrac{1}{2}$. Experimental data for device No. 67 from Fig. 2-13a are plotted in the form of Eq. (2-29) in Fig. 2-13b. This figure also shows a straight-line plot, the slope being a predicted $+\tfrac{1}{2}$.

The utility of Fig. 2-13a lies in the fact that all the constants of Eq. (2-28) can be found easily:

The slope of the line is equal to K_1 (notice that as the substrate doping level increases, K_1 increases).

The extrapolated intercept of the plot with the vertical axis (where $2\phi_F + V_{BG} = 0$) is V_{SS}.

The projection onto the vertical axis from the point on the curve where $V_{BG} = 0$ corresponds to V_{Ith}.

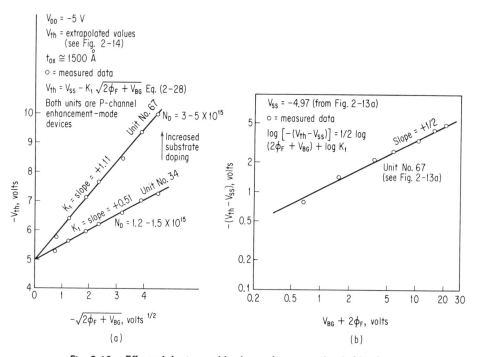

Fig. 2-13. Effect of doping and back-gate bias upon threshold voltage.

The value of ϕ_s at threshold can be found by curve fitting.

The difference between V_{th} and V_{SS} as measured on the vertical axis is equivalent to the intrinsic threshold voltage of Eq. (2-24).

An example using the data of Fig. 2-13a illustrates some of the device information obtainable. From the slope of the plotted line, K_1 is seen to be 1.11 for unit No. 67. Device No. 67 has an oxide thickness of approximately 1,500 Å. On Fig. 2-11a, a line is drawn from where $K_1 = 1.11$ to the 1,500 Å point. Projecting this line downward shows a substrate doping level of approximately 2×10^{15} (2.5 Ω-cm). (See graph in Appendix for resistivity vs. impurity density.) Projecting upward to the 1,500 Å point on V_{Ith} and over to the vertical axis gives $|V_{Ith}| = 0.88$ V. From Fig. 2-13a, V_{Ith} is read as $-5.75 + 4.95 = -0.8$ V. These two values agree quite well. The effective surface-state density for this device can be found from Eq. (2-26) as $Q_{SS}/q = -V_{SS}(\epsilon_{ox}/t_{ox})1/q$. Assuming $t_{ox} = 1,500$ Å and $\epsilon_{ox} = \frac{1}{3}$ pF/cm yields

$$\frac{Q_{SS}}{q} = -(-4.95)\frac{0.33 \times 10^{-12}}{1.5 \times 10^{-5} \times 1.6 \times 10^{-19}} \cong 6.9 \times 10^{11}/\text{cm}^2$$

Note that this value of effective surface-state density is of the same order of magnitude as that given in the discussion relating to Eq. (2-26).

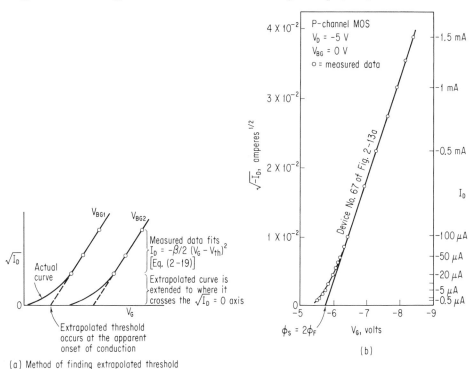

(a) Method of finding extrapolated threshold

Figure 2-14

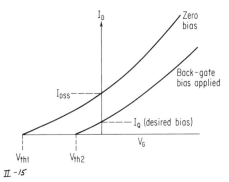

Fig. 2-15. Biasing through the use of the back-gate electrode.

II -15

The method used in Fig. 2-13 for determining the threshold voltage is illustrated in Fig. 2-14. Data are plotted which are known to fit a given equation. By extrapolating the data to a selected intercept, the extrapolated threshold is found. Threshold voltage found by this method would best be described by Eq. (2-28). (See also Sec. 3-1.) A plot of actual $\sqrt{I_D}$ vs. V_G data is shown in Fig. 2-14b.

An immediate application that comes to mind is the biasing of a depletion-mode device through back-gate control so that no low-impedance bias network is necessary on the front gate. This type of arrangement allows full use of the high input impedance of the MOS input.

Figure 2-15 shows the transfer curve for a hypothetical depletion-mode device. For zero bias, a drain current of I_{DSS} flows. The desired bias level is shown as I_Q. By applying a back-gate bias, the threshold voltage moves to the right from V_{th1} to V_{th2}. As a consequence, the output current drops to the desired value. This scheme can also be used in conjunction with a bias on the front gate, as seen in Fig. 2-12.

Equation (2-28) describes the threshold voltage as a function of the back-gate bias. It is often convenient to express V_{th} in terms of a fixed value (where $V_{BG} = 0$) plus a variable term, ΔV_{th}, which *is* a function of V_{BG}. The following analysis derives ΔV_{th} by starting with Eq. (2-28):

$$V_{th}(V_{BG}) = V_{SS} - K_1 \sqrt{2\phi_F + V_{BG}}$$

$K_1 \sqrt{2\phi_F}$ is added and subtracted to the right-hand side of the equation:

$$V_{th}(V_{BG}) = V_{SS} - K_1 \sqrt{2\phi_F} + K_1 \sqrt{2\phi_F} - K_1 \sqrt{2\phi_F + V_{BG}}$$

$$\boxed{V_{th}(V_{BG}) = V_{th} + \Delta V_{th}} \tag{2-30}$$

so that

$$\Delta V_{th} \equiv -K_1(\sqrt{2\phi_F + V_{BG}} - \sqrt{2\phi_F})^* \tag{2-31}$$

This result is plotted in Fig. 2-16 for various values of K_1 from Fig. 2-11a.

* For similar equations arrived at independently, see Ref. 17, p. 12, and Ref. 19, Eq. (7a).

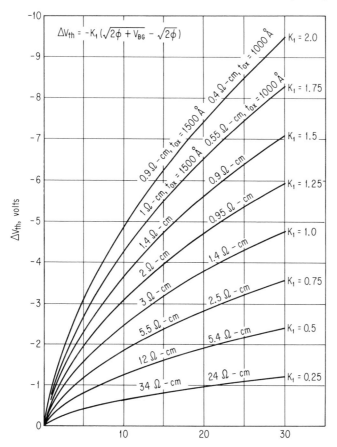

$$\Delta V_{th} = -K_1 \left(\sqrt{2\phi + V_{BG}} - \sqrt{2\phi} \right)$$

Fig. 2-16.   Threshold-voltage variation vs. back-gate bias.

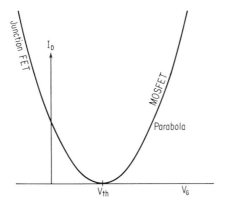

Figure 2-17

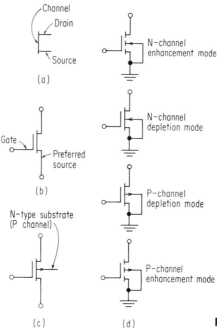

Fig. 2-18. MOS symbol notation.

Table 2-1 *, †

$$I_D = -\beta[(V_G - V_{th})V_D - \tfrac{1}{2}V_D{}^2]$$
    triode region, defined by $|V_D| < |V_G - V_{th}|$

$$I_D = -\frac{\beta}{2}(V_G - V_{th})^2$$
    saturation region, defined by $|V_D| \geq |V_G - V_{th}|$

---

1. N-channel device, valid for $(V_G - V_{th}) > 0$
    $\beta < 0$
    $V_D, I_D > 0$

  Depletion mode   Enhancement mode
    $V_{th} < 0$ $V_{th} > 0$
                                             $V_G > 0$

2. P-channel device, valid for $(V_G - V_{th}) < 0$
    $\beta > 0$
    $V_D, I_D < 0$

  Depletion mode   Enhancement mode
    $V_{th} > 0$ $V_{th} < 0$
                                           $V_G < 0$

---

  * See also Fig. (2-5).

  † The equations in this table assume that $R_s = R_d = 0$. For the case of finite parasitic source and drain resistance, see Eqs. (2-16a) and (2-16b).

2-4 SUMMARY OF MOS EQUATIONS AND DEVICE SYMBOLOGY

1. Equations. Because of the different polarities and combinations that are possible with MOSFET structures, Table 2-1 has been compiled to eliminate sign confusion. The device equations listed in the table are found in Eqs. (2-16) and (2-19) of the text. The notation $V_D > 0$ of course means that the voltage is positive, while $V_D < 0$ denotes a negative number.

When employing the equation valid for the saturation region, care must be taken to avoid the pitfall of using the "wrong" half of the transfer curve. For an N-channel enhancement-mode device, Eq. (2-19) gives the output current as a function of the square of the input voltage. Plotting this equation with no restraints on the left half of the curve yields a parabola, as shown in Fig. 2-17. The portion of the curve to the left of V_{th} exists mathematically but does not exist in the MOS device itself. A junction field-effect device, because it has essentially the same square-law characteristic as the MOS, is also represented by a parabolic transfer curve. However, because of its internal workings, the junction FET would be restricted to the left half of the parabola.

2. Symbology. The basic symbols used to designate MOS devices in this book are found in Fig. 2-18. It is highly desirable to convey as much information as possible about the device from its schematic representation. Whenever possible, these general rules should be followed in drawing the MOSFET:

1. The source and drain function as ohmic contacts and thus are shown drawn at right angles to the channel (Fig. 2-18a).
2. Enhancement-type devices do not have current flow for zero gate bias and are represented by a broken drain line. Depletion devices with initial conduction at zero gate bias are represented by a solid line.
3. MOS gates, because they are insulated and not a P-N junction, are shown with an L-shaped symbol where one side is parallel to the channel (Fig. 2-18b). The corner of the gate is placed opposite the preferred source.
4. The substrate, or back gate, is shown as a nonemitting diode and, as such, is drawn perpendicular to the channel. The direction of the arrowhead indicates the type of conductivity of the substrate (Fig. 2-18c).

BIBLIOGRAPHY

1. Ihantola, H. K. J.: Design Theory of a Surface-field-effect Transistor, *Stanford Electron. Lab. Rept.* 1661-1, September, 1961.
——— and J. L. Moll: Design Theory of a Surface-field-effect Transistor, *Solid-state Electron.*, vol. 7, pp. 423–430, 1964.
2. Grove, A. S., B. E. Deal, E. H. Snow, and C. T. Sah: Investigation of Thermally Oxidized Silicon Surfaces Using Metal-oxide-semiconductor Structures, *Solid-state Electron.*, vol. 8, pp. 145–163, 1965.
3. Sah, C. T.: Characteristics of the Metal-oxide-semiconductor Transistors, *IEEE Trans. Electron. Devices*, vol. ED-11, pp. 324–345, July, 1964.

4. Terman, L. M.: An Investigation of Surface States at a Silicon/Silicon Oxide Interface Employing Metal-oxide-silicon Diodes, *Solid-state Electron.*, vol. 5, pp. 285–299, 1962.
5. Heiman, F. P., and G. Warfield: The Effects of Oxide Traps on the MOS Capacitance, *IEEE Trans. Electron. Devices*, vol. ED-12, no. 4, pp. 167–178, April, 1965.
6. Zaininger, K. H., and G. Warfield: Limitations of the MOS Capacitance Method for the Determination of Semiconductor Surface Properties, *IEEE Trans. Electron. Devices*, vol. ED-12, no. 4, pp. 179–192, April, 1965.
7. Brown, W. L.: N-type Surface Conductivity on P-type Germanium, *Phys. Rev.*, vol. 91, no. 3, pp. 518–527, Aug. 1, 1953.
8. Goldberg, Colman: Space Charge Regions in Semiconductors, *Solid-state Electron.*, vol. 7, pp. 593–609, 1964.
9. Garrett, C. G. B., and W. H. Brattain: Physical Theory of Semiconductor Surfaces, *Phys. Rev.*, vol. 99, no. 2, pp. 376–387, July 15, 1955.
10. Greene, R., and T. Soldano: Increasing the Accuracy of MOS Calculations, *Proc. IEEE*, pp. 1241–1242, September, 1965.
11. Hofstein, S. R., and G. Warfield: Carrier Mobility and Current Saturation in the MOS Transistor, *IEEE Trans. Electron. Devices*, vol. ED-12, no. 3, pp. 129–138, March, 1965.
12. Reddi, V. G. K., and C. T. Sah: Source to Drain Resistance beyond Pinch-off in Metal-oxide-semiconductor Transistors (MOST), *IEEE Trans. Electron. Devices*, vol. ED-12, no. 3, pp. 139–141, March, 1965.
13. Early, J.: Effects of Space Charge Layer Widening in Junction Transistors, *Proc. IRE*, vol. 40, pp. 1401–1406, November, 1952.
14. Hofstein, S. R., and F. P. Heiman: The Silicon Insulated-gate Field-effect Transistor, *Proc. IEEE*, pp. 1190–1202, September, 1963.
15. Lindmayer, J., and C. Y. Wrigley: "Fundamentals of Semiconductor Devices," p. 345, D. Van Nostrand Company, Inc., Princeton, N.J., 1965.
16. Heiman, F. P., and H. S. Miller: Temperature Dependence of N-type MOS Transistors, *IEEE Trans. Electron. Devices*, vol. ED-12, no. 3, pp. 142–148, March, 1965.
17. Rapp, A. K., and R. S. Silver: Integrated Logic Nets, *Final Rept.*, Air Force Contract AF19-(604)-8836, July 31, 1964, RCA Laboratories, Princeton, N.J.
18. Sah, C. T., and H. C. Pao: The Effects of Fixed Bulk Charge on the Characteristics of Metal-oxide-semiconductor Transistors, *IEEE Trans. Electron. Devices*, vol. ED-13, no. 4, pp. 393–409, April, 1966.
19. Vadasz, L.: The Use of MOS Structure for the Design of High Value Resistors in Monolithic Integrated Circuits, *IEEE Trans. Electron. Devices*, vol. ED-13, no. 5, pp. 459–465, May, 1966.

GENERAL REFERENCES

Eimbinder, J.: The Field-effect Transistor: A "Curiosity" Comes of Age, *Electronics*, pp. 46–49, Nov. 30, 1964.
Grove, A. S., P. Lamond, et al.: Stable MOS Transistors, *Electro-technol.*, New York, pp. 40–43, December, 1965.
———, E. H. Snow, B. E. Deal, and C. T. Sah: Simple Physical Model for the Space-charge Capacitance of Metal-oxide-semiconductor Structures, *J. Appl. Phys.*, vol. 35, pp. 2458–2460, August, 1964.

Hall, R., and J. P. White: Surface Capacity of Oxide Coated Semiconductors, *Solid-state Electron.*, vol. 8, pp. 211–226, March, 1965.

Heiman, F. P., and S. Hofstein: Metal-oxide-semiconductor Field-effect Transistors, *Electronics*, pp. 50–61, Nov. 30, 1964.

—— and G. Warfield: The Effects of Oxide Traps on the MOS Capacitance, *IEEE Trans.*, vol. ED-12, pp. 167–178, April, 1965.

IEEE Trans., vol. ED-12, whole issue, March, 1965.

Pao, H. C., and C. T. Sah: Effects of Diffusion Current on Characteristics of Metal-oxide (Insulator)-semiconductor Transistors, *Solid-state Electron.*, vol. 9, pp. 927–937, October, 1966.

Tombs, N. C., et al.: A New Insulated-gate Silicon Transistor, *Proc. IEEE, Correspondence*, pp. 87–88, January, 1966.

Weimer, P. K.: The TFT—A New Thin Film Transistor, *Proc. IRE*, pp. 1462–1469, June, 1962.

Zaininger, K. H., and G. Warfield: Limitations of the MOS Capacitance Method for the Determination of Semiconductor Surface Properties, *IEEE Trans.*, vol. ED-12. pp. 179–192, April, 1965.

3

MOS Characteristics and
Equation Interrelationships

In Chap. 2, the pertinent device equations for the MOS were developed. The purpose of this chapter is threefold: First, experimental data will verify the device equations; second, from these relationships additional expressions will be obtained that will be useful in circuit design; and third, a device model will be developed to account for the variation of mobility as a function of gate voltage.

3-1 SATURATION REGION

1. I_D vs. V_G. Table 2-1 gives the relationship between the output drain current and the input gate voltage for a device in the saturation region:

$$I_D = -\frac{\beta}{2}(V_G - V_P)^2 \bigg|_{V_D} \tag{3-1}$$

[Notice that Eq. (3-1) is valid only for a fixed drain voltage. If the drain voltage is varied, then Eqs. (2-20) and (2-21) must be considered]. As mentioned earlier, Eq. (3-1) yields a square-law transfer curve. For various applications (especially where odd harmonics cannot be tolerated in the output), this unique field-effect feature makes both MOS and junction FET's very desirable devices.*

The square-law characteristic can be examined more closely by plotting $\sqrt{|I_D|}$ vs. V_G. Taking the square root of both sides of Eq. (3-1) and rearranging the terms yield

$$\sqrt{|I_D|} = +\sqrt{\frac{\beta}{2}}\, V_G - \sqrt{\frac{\beta}{2}}\, V_P \tag{3-2}$$

Equation (3-1) written in this manner is now in the same form as the general

---

* For an excellent application utilizing the square-law behavior of the junction FET, see Ref. 1, pp. 80 to 84, in the Bibliography at the end of the chapter.

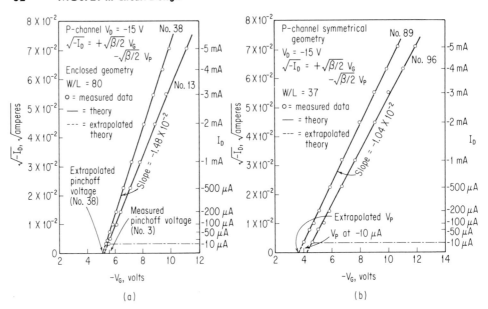

Fig. 3-1. Straight-line relationship between $\sqrt{I_D}$ and gate voltage.

expression for a straight line:

$$y = mx + b$$

where m = slope

b = constant

If, indeed, the square-law relationship for the MOS is valid, then data placed into Eq. (3-2) will plot as a straight line.   As seen earlier, in Fig. 2-16, and now in Fig. 3-1, a straight line is, in fact, observed over a wide current range and for devices of various geometries.*   Thus Eq. (3-2) is seen to be a reasonable representation or model of the MOS in the saturation region.   Below about 10 μA, the measured data depart from the predicted behavior, as noted in Sec. 2-3.   Figure 3-1a refers to a device type that has the drain region enclosed or surrounded by the source, a channel length L of 0.5 mil, and a channel width of 40 mils.   Figure 3-1b is for units with the drain and source in parallel, with a channel length of 0.5 mil, and with a channel width of 18.5 mils.

Certain useful information regarding the device is obtained by plotting $\sqrt{|I_D|}$ vs. V_G.   Notice that if $\sqrt{|I_D|}$ is set equal to zero in Eq. (3-2), the gate voltage then becomes equal to the pinchoff voltage.   By extrapolating the curve shown in Fig. 3-1 down to where $\sqrt{|I_D|} = 0$, the extrapolated threshold or pinchoff

---

* This fact is also observed by Sah (Ref. 3 in the Bibliography for Chap. 2), Heiman and Hofstein,[2] and Harrap et al.[3]   (Superscript numbers indicate works listed in the Bibliography at the end of this chapter.)

voltage can be read directly from the V_G axis (see also Fig. 2-1b). The device constant β can also be found from this type of plot by noting that the slope of the line is $\sqrt{\beta/2}$.

There are two methods for measuring a threshold or pinchoff voltage: The first is that described above, or by extrapolation; the second involves measuring the gate voltage required to produce some small drain current (usually 10 μA). This second method is generally used for the pinchoff voltage specified on the manufacturer's data sheet. From the curves of Fig. 3-1, it can be seen that the measured value at -10 μA approximates fairly well (but not exactly) the extrapolated value.

Once V_P is known, the device constant β can be found either from the slope of the curve in Fig. 3-1 or directly from Eq. (3-1). By knowing both β and V_P, the drain current can then be calculated for any given gate voltage. The following example will illustrate this point.

From Fig. 3-1a, device No. 38 is seen to have a theoretical pinchoff voltage of -5.1 V. Placing this into Eq. (3-1) together with one other point from the graph (-1 mA, -7.2 V) yields

$$\beta = -\frac{2I_D}{(V_G - V_P)^2} \quad \text{mhos/V}$$

$$\beta = -\frac{-2 \times 10^{-3}}{(-7.2 \text{ V} + 5.1)^2} = +454 \text{ } \mu\text{mhos/V}$$

Using Eq. (3-1), V_P and β will predict the current for a gate voltage of -9.4 V:

$$I_D = -\frac{454 \text{ } \mu\text{mhos/V}}{2}(-9.4 + 5.1)^2 - 227 \times 10^{-6}(18.5) = -4.2 \text{ mA}$$

The predicted value of I_D is -4.2 mA, as opposed to the -4.0 mA read from the graph—a difference of only 5 percent.

2. Transconductance. The "gain" parameter of the MOSFET is the forward-transfer–conductance ratio. This expresses the output-current variation for an input-voltage variation. Common symbols for transconductance are Y_{fs} and g_m. Typical values of g_m for discrete devices range from 1,000 to 2,000 μmhos, while devices used in integrated form are generally an order of magnitude lower. A device exhibiting a g_m of 1,000 μmhos means that the output current is related to the input voltage in the ratio of 1 mA/V.

Transconductance is defined as the ratio of a small variation in the drain current to the variation in gate voltage which produces the current. Mathematically, this can be written as

$$g_m = \frac{\partial(I_D)}{\partial V_G}\bigg|_{V_D}$$

Differentiating Eq. (3-1) yields

$$\frac{\partial(I_D)}{\partial V_G} = \frac{\partial\left[-\dfrac{\beta}{2}(V_G - V_P)^2\right]}{\partial V_G} = \boxed{g_m = -\beta(V_G - V_P)} \qquad (3\text{-}3)$$

(g_m is a positive number.)   Notice that g_m is the product of the device constant β and the difference between the gate and the pinchoff voltage.   Thus g_m can be increased in one of two ways.   First, the gate bias V_G may be increased (this is available to the circuit designer as well as to the device designer), or second, the geometry of the device may be varied (this procedure is generally available only to the device designer).   Since

$$\beta = \frac{\mu \epsilon_{ox}}{t_{ox}} \frac{W}{L}$$

a simple geometrical variation in the width-to-length ratio will vary the g_m of the device to any desired value.

The first method suggested for increasing g_m, that of increasing V_G, can be verified by placing measured data into Eq. (3-3) and then plotting the result. This is shown in Fig. 3-2.   Equation (3-3) says that g_m vs. $-(V_G - V_P)$ should plot out to a straight line, which, in fact, it does.   However, the graph shows that for $-(V_G - V_P) = 0$ V, the devices still exhibit some finite gain.   This is not reasonable and does not agree with Eq. (3-3).   The apparent anomaly comes from the fact that the pinchoff voltage used to calculate the horizontal axis was the measured value ($I_D = -10$ μA) and consequently varied from the extrapolated value.   Extrapolating the curves back to the horizontal axis gives an intersection of from 0.2 to 0.4 V to the left of the origin.   This is just about the

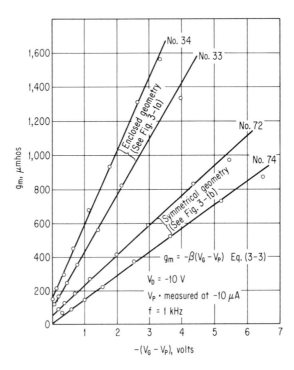

Fig. 3-2. g_m vs. $-(V_G - V_P)$.

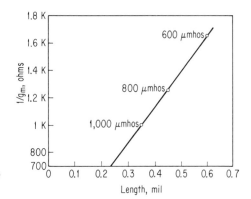

Fig. 3-3. $1/g_m$ vs. channel length. (*After Rapp and Silver, Ref. 17 for Chap. 2.*)

difference between the extrapolated and measured threshold voltages (see Fig. 3-1). Since the magnitude of the measured threshold is greater than the extrapolated value, the curves were shifted to the left by the amount of the difference. Adding the difference back into each point on the curves will, of course, replot each curve from the origin. This is an excellent example of the pitfall that may be encountered when the "measured" value of threshold is used.

The second method for increasing g_m suggested that the width-to-length ratio be increased. Transconductance as a function of the W/L ratio can be seen implicitly from the information contained in Fig. 3-1. Here the equation for the straight line shows that the slope is proportioned to $\sqrt{\beta}$. Thus the ratio of the slopes of device No. 38 to those of device No. 89 should equal the square root of their W/L ratios. This can be expressed as

$$\frac{\text{Slope (No. 38)}}{\text{Slope (No. 89)}} = \sqrt{\frac{W/L \text{ (No. 38)}}{W/L \text{ (No. 89)}}}$$

The values are read from the graph:

$$\frac{-1.47}{-1.02} = \sqrt{\frac{80}{37}}$$
$$1.44 \cong 1.47$$

It is seen that the measured ratio is quite close to the calculated ratio. Thus one can assume that the g_m can be set by a simple variation in the W/L ratio.

Further verification of the fact that g_m varies directly as the W/L ratio can be found by examining Fig. 3-3. Here data taken by Rapp and Silver* show that $1/g_m$ vs. L plots as a straight line. Equation (3-3) predicts this result when written as

$$\frac{1}{g_m} = L \frac{1}{W} K \frac{1}{V_G - V_P}$$

where K = constant.

Equation (3-3) neglects the effect that the drain voltage has upon the drain

* Reference 17 in the Bibliography for Chap. 2.

current.   The finite slope of the output characteristics in saturation, as described in Sec. 2-2, tends to spread the gate-voltage curves, so that g_m increases as the drain voltage is increased.   This is readily seen from Fig. 2-6, where the increment of drain current between $V_G = -10$ V and $V_G = -11$ V is greater at $V_D = -10$ V than at $V_D = -5$ V.

An expression can be derived for g_m as a function of drain current.   Combining the equations for g_m and I_D yields $g_m = f(I_D)$,

$$g_m = -\beta(V_G - V_P)$$

$$I_D = -\frac{\beta}{2}(V_G - V_P)^2$$

$$g_m = \sqrt{2|\beta|}\ \sqrt{|I_D|} \tag{3-4}$$

Equation (3-4) plots as a straight line on log-log paper with a slope of $+\frac{1}{2}$, and measured data should plot accordingly if Eqs. (3-3) and (3-4) are valid representations of the gain of the MOS transistor.   The log of g_m vs. the log of I_D for experimental data on a P-channel device does, in fact, yield a straight line, as shown in Figure 3-4.   Thus Eqs. (3-3) and (3-4) are seen to be reasonable mathematical models for an MOS in saturation with a fixed drain voltage.

g_m as Modified by a Source Resistance.   The gain of most semiconductor devices varies from device to device, varies with temperature, and in fact, varies with the change of almost any condition.   The g_m of a MOS transistor is no exception. Often it is desirable to stabilize the gain.   A common but effective method for stabilization is the addition of a source resistance R_S to provide local negative feedback (source degeneration).   In calculating the effect that R_S has upon g_m, it

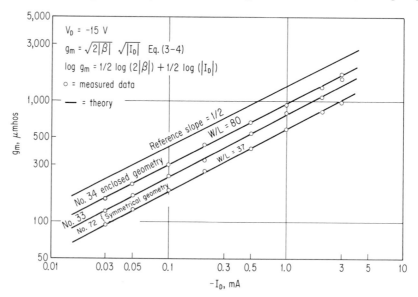

Fig. 3-4.   Log g_m vs. log I_D.

will be constructive to compare the percentages of variation between the circuit gain with and without a source resistance. The analysis will proceed as follows:

1. Write the d-c circuit equations.
2. Plug in the mathematical model derived earlier for the device.
3. Solve for v_{in} as a function of I_D.
4. Define a circuit transconductance g'_m as $g'_m = \partial I_D / \partial v_{in}$, as opposed to the device g_m, which equals $\partial I_D / \partial V_G$. (Notice that v_{in} and V_G are not equal.)
5. Take the derivative of the d-c equation and solve for g'_m as a function of g_m and R_S.

Figure 3-5 shows the circuit used in this derivation. Writing the input-loop equation yields

$$v_{in} = V_G + I_D R_S \tag{3-5}$$

Setting Eq. (3-3) equal to Eq. (3-4), solving for V_G, and substituting into Eq. (3-5) give

$$v_{in} = \sqrt{\frac{2}{\beta}(-I_D)} - V_P + I_D R_S$$

Now

$$\frac{1}{g'_m} \equiv \frac{\partial v_{in}}{\partial I_D} = \left(\frac{2}{\beta}\right)^{\frac{1}{2}} \frac{1}{2(-I_D)^{\frac{1}{2}}}$$

Rearranging terms and using the relations

$$I_D = -\frac{\beta}{2}(V_G - V_P)^2$$

$$V_G - V_P = -\sqrt{\frac{2(-I_D)}{\beta}}$$

yield

$$g'_m = \frac{g_m}{1 + R_S g_m} \tag{3-6}$$

The limits on g'_m can be found by letting R_S approach 0 and $R_S g_m$ become larger than 1. As R_S approaches 0, the circuit transconductance approaches that of the device. As $g_m R_S$ becomes large, g'_m approaches $1/R_S$. Thus, when $g_m R_S$ is large, the circuit transconductance becomes independent of the device parameters and is fixed at $1/R_S$.

With a finite source, resistance variations in g_m produce variations in g'_m. The variations in g'_m can be made as small as desired for a given Δg_m by using sufficient

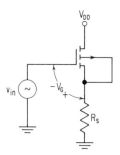

Figure 3-5

source degenerative feedback. The following analysis considers large trans-conductance variations, Δg_m, rather than incremental changes, dg_m, because in practical cases gains vary as much as 2:1 to 5:1. Circuit transconductance variations can be written as

$$\overline{g'_m} - \underline{g'_m} \equiv \Delta g'_m = \frac{\overline{g_m}}{1 + R_s\overline{g_m}} - \frac{\underline{g_m}}{1 + R_s\underline{g_m}}$$

where $\overline{g_m}$ and $\underline{g_m}$ represent the maximum and minimum values, respectively.

Dividing through by $\underline{g'_m}$ to find the percentage of variation with respect to the minimum transconductance yields

$$\frac{\Delta g'_m}{\underline{g'_m}} = \frac{1 + R_s\underline{g_m}}{1 + R_s\overline{g_m}} \frac{\overline{g_m}}{\underline{g_m}} - 1$$

Rearranging terms gives the final result:

$$\frac{\Delta g'_m}{\underline{g'_m}} \times 100 = \frac{1}{1 + R_s\overline{g_m}} \frac{\Delta g_m}{\underline{g_m}} \times 100$$

which can also be written as

$$\frac{\Delta g'_m}{\underline{g'_m}} \times 100 = \frac{\overline{g'_m}}{\overline{g_m}} \frac{\Delta g_m}{\underline{g_m}} \times 100 \qquad (3\text{-}7)$$

Equation (3-7) says that the percentage of variation in g'_m is equal to the percentage of variation in g_m times the factor $1/(1 + R_s\overline{g_m})$ (or $\overline{g'_m}/\overline{g_m}$). The ratio will always be ≤ 1, so that the circuit gain variation ($\Delta g'_m$) can be made as small as desired for a given Δg_m by making the factor $1/(1 + R_s\overline{g_m})$ small.

An example will illustrate the above. Assume a device has a 2:1 transconductance spread—minimum g_m at a given bias point is 2,000 μmhos, but it can vary to a maximum of 4,000 μmhos. Now assume the specific circuit requires a g_m of only 800 μmhos for proper operation. How much source resistance can be added, and how much does this reduce the 2:1 variation in g_m? Equation (3-6) is used to solve for R_S:

$$800 \times 10^{-6} = \frac{2{,}000 \times 10^{-6}}{1 + R_s \times 2{,}000 \times 10^{-6}} \qquad R_S = 750 \ \Omega$$

Thus, the addition of 750 Ω in the source will result in an effective g'_m of 800 μmhos.

Equation (3-7) gives the percentage of variation in g'_m for a 100 percent variation in g_m:

$$\frac{\Delta g'_m \times 100}{\underline{g'_m}} = \frac{1{,}000 \ \mu\text{mhos}}{4{,}000 \ \mu\text{mhos}} \frac{(4{,}000 - 2{,}000) \ \mu\text{mhos}}{2{,}000 \ \mu\text{mhos}} \times 100$$

$$\frac{\Delta g'_m}{\underline{g'_m}} \times 100 = \frac{1}{4} \times 100 = 25\%$$

The example shows that as the device gain varies from 2,000 to 4,000 μmhos (100 percent), the effective gain with a 750-Ω source resistance varies only 25 percent, from 800 to 1,000 μmhos.

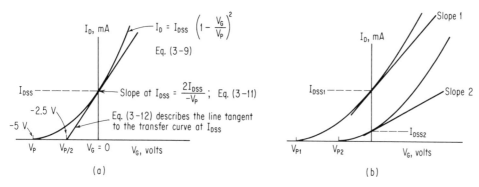

Figure 3-6

3. Additional Equation Interrelationships.   A typical transfer curve for a deple-tion-type N-channel device (either MOS or junction FET) is shown in Fig. 3-6a.* This curve has an ideal built-in reference point—the drain current at zero gate voltage.   It is useful to express some of the device equations in terms of this easily measured current, which is specified as either I_{DSS} or $I_{D(ON)}$.

At the present time, N-channel devices are the only generally available deple-tion-mode MOSFET's.   However, this discussion is general and applies to P-channel as well as N-channel devices with the appropriate change in signs.

From Table 2-1, the equation for the saturation drain current is given as

$$I_D = \frac{-\beta}{2}(V_G - V_P)^2 \tag{3-1}$$

Setting the gate voltage to zero yields an expression for zero-bias current flow:

$$I_D\Big|_{V_G=0} \equiv I_{DSS} = \frac{-\beta}{2}(-V_P)^2 \tag{3-8}$$

This current is a function of both the geometry of the device and the pinchoff voltage.

Often it is desirable to normalize the output current as defined in Eq. (3-1). By factoring out $-V_P$, the equation can be normalized to I_{DSS} and V_P:

$$I_D = \frac{-\beta}{2}(-V_P)^2\left(1 - \frac{V_G}{V_P}\right)^2$$

$$= I_{DSS}\left(1 - \frac{V_G}{V_P}\right)^2$$

$$\frac{I_D}{I_{DSS}} = \left(1 - \frac{V_G}{V_P}\right)^2 \tag{3-9}$$

* See Fig. 2-5 for transfer curves of both N-channel and P-channel enhancement and depletion devices.

From Eq. (3-3), g_m has been defined as a change in drain current with respect to a change in gate voltage. Looking at Fig. 3-6, one sees that this definition is just the slope of the transfer curve. An interesting relationship can be seen by finding the slope at zero bias voltage:

$$\frac{\partial I_D}{\partial V_G}\bigg|_{V_G=0} \equiv g_{mo} = \frac{-\beta}{2}\,2(-V_P) \tag{3-10}$$

Multiplying top and bottom by $-V_P$ yields

$$g_{mo} = \frac{2I_{DSS}}{-V_P} \tag{3-11}$$

(V_P is negative in Fig. 3-6, giving a positive value to g_{mo}. This is in agreement with the positive slope drawn on the figure.) Equation (3-11) states that three of the basic FET parameters are inextricably tied together. By specifying any two of the parameters, the third is automatically fixed. One can obtain an idea of why these parameters are interrelated in such a way as to be mutually dependent by observing variations in the transfer curve as shown in Fig. 3-6b. Increasing I_{DSS} while maintaining a constant V_P results in a steeper slope, producing a higher g_m. Moving the transfer curve to the right results in an increase in V_P, a lowering of I_{DSS}, and a reduced slope. Other variations to illustrate the relationship between V_P, I_{DSS}, and g_{mo} are obviously possible.

The intercept of the line tangent to the transfer curve at I_{DSS} with the V_G axis occurs at exactly $V_P/2$ for a square-law device. This may be seen by writing the equation for the tangent line in the form of $y = mx + b$:

$$I_D = \frac{2I_{DSS}}{-V_P}\,V_G + I_{DSS} \tag{3-12}$$

The intercept is found when $I_D = 0$. Setting Eq. (3-12) to equal zero yields

$$\frac{2I_{DSS}}{V_P}\,V_G = I_{DSS}$$

$$V_G = \frac{V_P}{2} \tag{3-13}$$

If $V_P = -5$ V, then the tangent line would intercept the V_G axis at -2.5 V. This fact is shown graphically in Fig. 3-6a.

4. I_D, g_{mBG} as a Function of V_{BG}. In Sec. 2-3, the idea was presented that the back gate of the MOS, as well as the front gate, can modulate the drain current. If the equation for pinchoff voltage is substituted into the expression for the drain current at saturation, then an expression is developed which relates I_D to the back-gate bias. Combining Eqs. (2-19) and (2-28) yields the following equation for a P-channel device:

$$I_D = -\frac{\beta}{2}\,[K_1\sqrt{\phi_s + V_{BG}} + (V_G - V_{SS})]^2 \tag{3-14}$$

This equation has the same form as the square-law equation (2-19).   However, notice that the effective gate voltage is now $K_1 \sqrt{\phi_s + V_{BG}}$ and that the new pinchoff voltage is $V_G - V_{SS}$.   If the square root of Eq. (3-14) is plotted, a straight line should result.   This is illustrated in Fig. 3-7, where data plotted in this form verify the equation.   In the plot, a value of 0.8 V was selected for the surface potential ϕ_s.   This value was used because, considering the current levels involved in the measurement $|I_D| \gg 20$ μA, the surface tends to be heavily inverted. Figure 2-16 shows that for a typical case, the bands are bent somewhat greater than $2\phi_F (\phi_s > 2\phi_F)$ when $|I_D| > 20$ μA.

A question that occurs when the statement is made that the back gate can control the device current is: Just how effective is its control—or, in other words, what is the gain?   The gain or transconductance due to the back gate (specified as g_{mBG}) is completely analogous to the front-gate transconductance:

$$g_{mBG} = \left. \frac{\partial I_D}{\partial V_{BG}} \right|_{\text{for a constant } V_D \text{ and } V_G} \qquad (3\text{-}15)$$

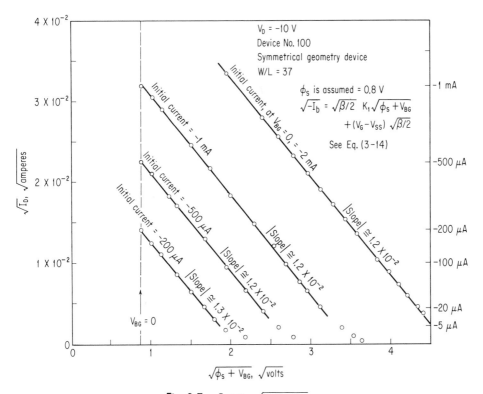

Fig. 3-7. I_D vs. $\sqrt{\phi_s + V_{BG}}$.

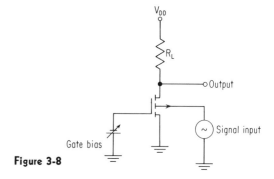

Gate bias

Figure 3-8

Equation (3-14) can be differentiated as specified by Eq. (3-15) to give

$$g_{mBG} = - \frac{(\beta/2)(V_G + K_1 \sqrt{\phi_s + V_{BG}} - V_{SS})K_1}{(\phi_s + V_{BG})^{1/2}} \tag{3-16}$$

While Eq. (3-16) gives g_{m2} as a function of the back-gate bias, it is a bit bulky to handle for our purposes here. Instead, a simpler case will be examined, i.e., where the d-c value of the back-gate bias is set equal to zero. This situation is depicted in Fig. 3-8.

Letting $V_{BG} = 0$ yields

$$g_{mBG} = - \frac{\beta[V_G - (K_1 \sqrt{\phi_s} + V_{SS})]K_1}{2 \sqrt{\phi_s}} \tag{3-17}$$

Rearranging the terms gives

$$g_{mBG} = - \beta(V_G - V_P) \frac{K_1}{2 \sqrt{\phi_s}} \tag{3-18}$$

Equation (3-18) states that the transconductance of the back gate is equal to the front-gate gain $[-\beta(V_G - V_P)$, Eq. (3-3)] times the factor $K_1/2 \sqrt{\phi}$, which typically ranges from 0.5 to 1.0. Equation (3-18) should plot fairly close to a straight line. The first term $[-\beta(V_G - V_P)]$ has already been shown in Fig. 3-3 to be a straight-line relationship. $K_1/2 \sqrt{\phi_s}$ approximates a constant only over a limited current range because of the fact that ϕ_s varies slightly as the current is varied. An interesting discussion on the operation of four-terminal junction FET's can be found in Ref. 4.

3-2 TRIODE REGION

In this section, operation will be restricted to the nonsaturated, or triode, region. Because of the strong influence the drain voltage has upon the drain current in this region, the MOSFET can be regarded as a two-terminal resistor between source and drain whose resistance can be modulated by the gate voltage. Four characteristics of the MOS make it ideal for use as an a-c or d-c small-signal switch:

1. Low values of channel resistance (100 to 500 Ω) obtainable.
2. Ability to modulate this resistance from high to low with the gate voltage.*
3. Zero offset voltage.
4. Symmetrical operation about the origin in a positive and negative direction.

1. Drain Resistance.   Table 2-1 lists the device equation for an enhancement P-channel device as

$$I_D = -\beta[(V_G - V_{th})V_D - \tfrac{1}{2}V_D{}^2] \qquad (3\text{-}19)$$

The drain conductance in the triode region can be found by differentiating Eq. (3-19) as follows:

$$\frac{dI_D}{dV_D} = -\beta(V_G - V_P) + \beta V_D \qquad (3\text{-}20)$$

The first term on the right side of Eq. (3-20) is positive, while the second term is always negative (for a P channel). However, in the triode region, $|V_D| < |V_G - V_P|$, so that Eq. (3-20) yields a positive output conductance.

The output resistance close to the origin can be found from Eq. (3-20) by letting V_D approach 0:

$$r_{dt} = \frac{1}{-\beta(V_G - V_P)} \qquad (3\text{-}21)$$

(The symbol r_{dt} is used to signify the incremental drain resistance in the triode region, as opposed to r_{ds} which signifies the drain resistance in the saturation region.

Close to the origin ($|V_D| \leq 0.15$ V), the incremental impedance equals the static impedance since the curves are straight and pass through the origin. In this range, Eq. (3-21) is valid for either a-c or d-c voltages. Figure 3-9 shows MOS characteristics near the origin for a device from the TIXS11 family. Note that the V-I characteristics are linear, pass through, and are reasonably sym-

* For a discussion of the junction FET as a voltage-controlled resistor, see Refs. 5 and 6.

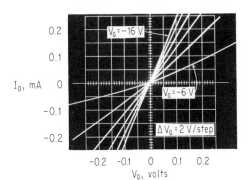

Fig. 3-9. I_D vs. V_D, **triode-region ON resist-ance.**

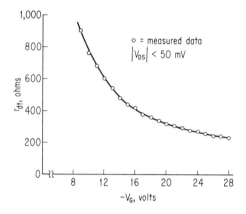

Fig. 3-10. r_{dt} vs. V_G.

metrical about the origin for both a positive and a negative drain voltage. A plot of r_{dt} as a function of gate voltage is shown in Fig. 3-10.

Note that the reciprocal of the channel resistance given in Eq. (3-21) is equal to the transconductance of the device at saturation as given by Eq. (3-3). Thus the ON resistance is directly related to the device gain. The device with a transconductance of 1,000 μmhos (in saturation) is characterized with an ON channel

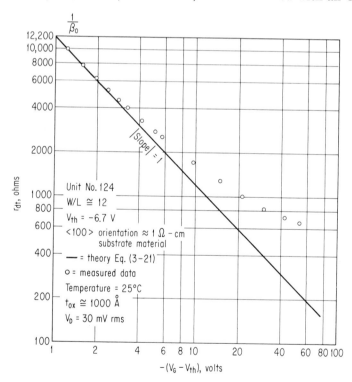

Fig. 3-11. Triode-region drain resistance vs. $V_G - V_{th}$.

resistance of $1,000 \ \Omega \ (V_D = 0)$, while a g_m of $4,000 \ \mu$mhos corresponds to a 250-Ω device.

It is informative to plot measured data in the form of Eq. (3-21) on log-log paper. A straight-line graph with a |slope| $= 1$ should result if the equation is a reasonable representation of the device. This plot is shown in Fig. 3-11. Notice that the measured data fit the unity-slope theory in the low-gate-voltage region but depart from unity slope at higher gate voltages. Thus it is obvious that some modification is necessary to account for the departure of r_{dt} from the simple theory at high gate voltages.

In Chap. 2, it was assumed that mobility was constant and independent of the applied gate voltage. This is not actually the case (although the approximation of constant mobility does give adequate predictability for circuit design in most cases), for mobility is a function of the gate voltage, as shown by Fang and Triebwasser,[7] Fowler et al.,[8] and Colman and Mize.[9] Including mobility variation in Eq. (3-21) results in a theory that fits the measured data. Equation (3-21) can thus be rewritten as

$$r_{dt} = \frac{1}{-\beta(V_G)(V_G - V_{th})} \tag{3-22}$$

where $\beta(V_G)$ indicates a dependence upon gate voltage.

2. g_m in the Triode Region. A dynamic illustration of the deleterious effect of mobility variation on device performance is offered by transconductance in the triode region. Transconductance in the triode region is defined in the same manner as in the saturation region; that is,

$$g_m \equiv \frac{\partial I_D}{\partial V_G}\bigg|_{V_D} \tag{3-23}$$

Differentiating Eq. (3-19) with respect to V_G yields

$$g_m = -\beta V_D \tag{3-24}$$

This equation shows that one would expect the g_m to be dependent only upon the drain voltage and to be independent of the d-c gate bias voltage. [This analysis also assumes mobility (μ_p) occurring in the β term to be independent of V_G (approximation No. 1).] Figure 3-12 is a scope photograph of g_m vs. V_G.

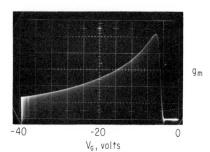

Fig. 3-12. Transconductance in triode region vs. gate voltage.

-40   -20   0

V_G, volts

g_m

Notice that g_m falls off rather drastically as the gate voltage exceeds 6 V in the negative direction. This variation in g_m, due to mobility reduction, can be predicted by the addition to Eq. (3-24) of a mobility term which is dependent upon V_G. Differentiating Eq. (3-19) with respect to V_G while considering gate-voltage dependence upon mobility yields[10]

$$g_m = -\beta(V_G)V_D - \left|\frac{\partial\mu(V_G)}{\partial V_G}\right|\left(\frac{\epsilon_{ox}}{t_{ox}}\frac{W}{L}\right)[(V_G - V_{th})V_D - \tfrac{1}{2}V_D{}^2] \quad (3\text{-}25)$$

where $\beta(V_G)$ and $\mu(V_G)$ indicate a dependency upon gate voltage. The second part of Eq. (3-25) contributes significantly to the g_m falloff because of the $\partial\mu(V_G)/\partial V_G$ term.

3-3   MOBILITY

The purpose of this section is to develop an empirical relation expressing mobility as a function of applied gate voltage.

Examination of Fig. 3-11 shows that the measured data depart from the straight-line theory by a constant amount of approximately 420 Ω. Subtracting 420 Ω from each datum point in the figure results in a straight-line relationship,

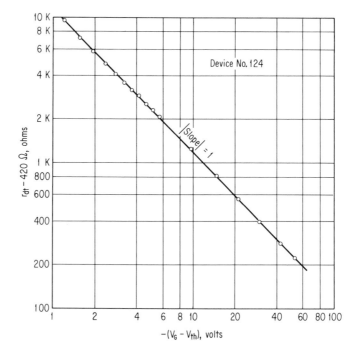

Fig. 3-13.   Measured data resulting in the model expressed by Eq. (3-26).

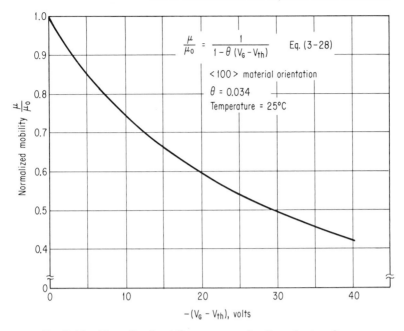

Fig. 3-14. Normalized mobility curve as a function of gate voltage.

as illustrated in Fig. 3-13. This result suggests a model or an equivalent circuit for a practical structure consisting of an "ideal" MOS, whose mobility μ_o is independent of gate voltage, in series with a constant resistance R. Such an equivalent circuit would yield the following expression for resistance in the triode region when $V_D \cong 0$:

$$r_{dt} = \frac{1}{-\beta_o(V_G - V_{th})} + R \quad *$$

(3-26)

where $R = 420\ \Omega$ for the device of Figs. 3-11 and 3-13 and β_o, the initial low-voltage value, $\cong 82\ \mu$mhos/V.

There are now two expressions for triode resistance—the theoretical result of Eq. (3-22) and the experimental relationship of Eq. (3-26). Equating these two expressions results in a simple empirical expression for $\beta(V_G)$ or $\mu(V_G)$ that predicts the shape of the mobility-vs.-gate-voltage curve:

$$\frac{1}{-\beta(V_G)(V_G - V_{th})} = \frac{1}{-\beta_o(V_G - V_{th})} + R$$

(3-27)

* When using this equation, it must be remembered that if an actual parasitic resistance *were* present in the transistor, it is possible that its effect would dominate so that the true mobility variation would not be seen.

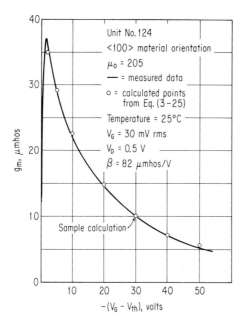

Fig. 3-15. g_m **falloff in the triode region showing good agreement between measured and calculated data.**

Rearranging Eq. (3-27) results in the expression

$$\frac{\beta(V_G)}{\beta_o} = \frac{\mu(V_G)}{\mu_o} = \frac{1}{1 - \theta(V_G - V_{th})} \qquad (3\text{-}28)$$

where $\theta = \beta_o R$.

Two constants are needed for the evaluation of Eq. (3-28): μ_o, which determines the magnitude of the mobility curve; and θ, which determines the shape of the curve. *Typical* values the author has seen for μ_0 and θ are tabulated below:*

	⟨100⟩ Material	⟨111⟩ Material
μ_o (cm²/V-sec)	180–200	240–250
θ (V⁻¹)	0.034	0.028

1 to 10 Ω-cm N-type material; temperature = 25°C.

The product of β_o and R from Fig. 3-11 gives $\theta = 0.0344$. A normalized mobility curve for ⟨100⟩ material is given in Fig. 3-14.

* Only two crystal planes have been considered in this discussion, the ⟨100⟩ and the ⟨111⟩. At the time of writing, information on other planes was not available.

If Eq. (3-28) is indeed a true representation of mobility as a function of gate voltage, then it should be able to predict the g_m falloff as described in Eq. (3-25). Excellent agreement between the theory and the measured data is illustrated in Fig. 3-15.

A sample calculation for g_m at -30 V will be carried out.   First, it is necessary to know $\partial \mu(V_G)/\partial V_G$.   From Eq. (3-28),

$$\frac{\partial \mu(V_G)}{\partial V_G} = \frac{\theta \mu_o}{[1 - \theta(V_G - V_{th})]^2} \tag{3-29}$$

For $\theta = +0.034$ V, $\mu_o = +205$ cm^2/V-sec, and $(V_G - V_{th}) = -30$ V,

$$\left| \frac{\partial \mu(V_G)}{\partial V_G} \right| = 1.72 \text{ cm}^2/\text{V}^2\text{-sec}$$

From Fig. 3-14, $\beta(V_G)$ at -30 V $= +0.495 \times 82$ μmhos/V $= +40.6$ μmhos/V.

$$\frac{\epsilon_{ox}}{t_{ox}} \frac{W}{L} = \frac{\beta_0}{\mu_0} = \frac{82 \times 10^{-6}}{205} = 0.4 \times 10^{-6}$$

From Eq. (3-25), $g_m = -40.6 \times 10^{-6}(-0.5) - 1.72(0.4 \times 10^{-6})[(-30)(-0.5) - 0.125] = (20.3 - 10.2) \times 10^{-6} = 10.1$ μmhos.   This point is designated as "sample calculation" in Fig. 3-15.

BIBLIOGRAPHY

1. Sevin, L. J., Jr.: "Field-effect Transistors," Texas Instruments Electronics Series, McGraw-Hill Book Company, New York, 1965.
2. Heiman, F. P., and S. R. Hofstein: Metal-oxide-semiconductor Field-effect Transistors, *Electronics*, pp. 50–61, Nov. 30, 1964.
3. Harrap, V., G. Pierson, H. Kuehler, and B. K. Lovelace: Researchers Turn to Germanium for a MOS Field-effect Transistor, *Electronics*, pp. 64–68, Nov. 30, 1964.
4. Latham, D. C., F. A. Lindholm, and D. J. Hamilton: Low-frequency Operation of Four-terminal Field-effect Transistors, *IEEE Trans. Electron. Devices*, vol. ED-11, no. 6, pp. 300–305, June, 1964.
5. Sherwin, J. S.: Voltage Controlled Resistor, *solid/state/design*, p. 12, August, 1965.
6. Martin, T. B.: Circuit Applications of the Field Effect Transistor, *Semicond. Prod.*, part I, pp. 33–39, February, 1962; part II, pp. 30–38, March, 1962.
7. Fang, F., and S. Triebwasser: Carrier Surface Scattering Inversion Layers, *IBM J. Res. Develop.*, vol. 8, no. 4, pp. 410–415, September, 1964.
8. Fowler, A. B., F. Fang, and F. Hochberg: Hall Measurements on Silicon Field Effect Transistor Structures, *IBM J. Res. Develop.*, vol. 8, no. 4, pp. 427–429, September, 1964.
9. Colman, D., and J. Mize: Hole Mobility in P-type Inversion Layers on Thermally Oxidized Silicon Surfaces, *IEEE Solid-state Device Res. Conf.*, Evanston, Ill., June, 1966.
10. Mize, J.: Personal Communication, Texas Instruments Inc.

4

Transient Response

The MOS is a fairly fast device, with intrinsic cutoff frequencies in the neighborhood of 1 GHz. In practical circuits, however, actual switching speeds are two to three orders of magnitude below the cutoff frequency. The cause of slower switching speeds, as may be expected, is stray circuit capacitance* that must be charged and discharged during the transition period. This section attempts to analyze this switching transient.

4-1 CUTOFF FREQUENCY

It may be recalled that a figure of merit for some voltage-controlled devices is the g_m/C_{in} ratio, where C_{in} is the input gate capacitance of the device. This ratio is roughly the 3-dB bandwidth of the device itself.† Equation (3-3) gives the saturation-region transconductance as $g_m = -\beta(V_G - V_P)$. The input capacitance is simply that of a parallel-plate capacitor of area A and thickness t_{ox}. Combining Eqs. (3-3) and

$$C_{\text{in}} = \frac{A\epsilon_{\text{ox}}}{t_{\text{ox}}} \tag{4-1}$$

for the g_m/C ratio yields‡

$$g_m/C = -\frac{\mu}{L^2}(V_G - V_P) \tag{4-2}$$

* *Stray capacitance* is defined as any capacitance unnecessary for actual device operation. Capacitance from drain to substrate need not exist for proper device operation. For this reason, drain capacitance is not included in the cutoff-frequency calculation. In theory, a device could be built in which stray capacitance was so small as to be negligible. If this were the case, the device would, in fact, be capable of operating at its theoretical cutoff frequency.

† For a thorough discussion of the MOS gain-bandwidth product and cutoff frequency, see Sah (Ref. 3 in the Bibliography for Chap. 2).

‡ See Ref. 15, p. 345, in the Bibliography for Chap. 2.

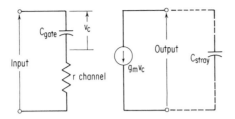

Fig. 4-1. Equivalent circuit for intrinsic cutoff-frequency analysis.

Notice that Eq. (4-2) is dependent only upon the length of the channel (direction of current flow) and not upon the width. Thus, increasing the channel width increases the capacitance as much as the transconductance, and no increase in cutoff frequency is achieved. Assuming a silicon device with a 0.2-mil channel length, a carrier mobility of 200 cm²/V-sec, and $V_G - V_P = -10$ V gives a device with a cutoff frequency of

$$f_c = \frac{g_m}{2\pi C} = 1.7 \text{ GHz} \tag{4-3}$$

(Note, however, that in a practical circuit where stray capacitance dominates, switching speeds are limited to 1 to 2 MHz.)

The physical significance of cutoff frequency can be seen by examining the simplified small-signal equivalent-circuit representation of the MOS shown in Fig. 4-1. Here the input circuit consists of the gate capacitance coupled to the channel resistance, which to a first approximation equals $1/g_m$. The very large gate resistance across the input is neglected. Notice that the control voltage is that which appears across the gate capacitance and is not necessarily equal to the input voltage. If zero stray capacitance together with a purely resistive load is assumed, then the output voltage will follow exactly the control voltage v_C. Supplying a step input gives the intrinsic rise time across C_{gate} as approximately 0.2 ns for the case of the previous example [Eq. (4-3)]. The output will follow v_C, resulting in a very fast theoretical switching time.

In contrast to the previous case, a significant amount of stray capacitance is assumed across the output. Again, application of a step input voltage results in a 0.2-ns rise time across C_{gate}. The current generator is turned on and supplies $g_m v_C$ amount of current within the 0.2 ns; however, it now takes a much longer time for the output voltage to follow. An example using typical IC values will

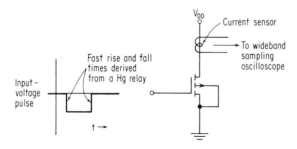

Fig. 4-2. High-speed circuit for determining cutoff-frequency time constant.

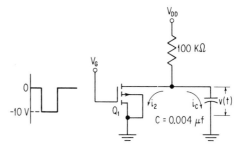

Fig. 4-3. Circuit used for common-source switching-time analysis.

give a rough approximation of practical switching speeds. Using the relation $tI = Cv$ and substituting in $C_{\text{stray}} = 1\,\text{pF}, v = 10\,\text{V}, I = 0.2\,\text{mA}$ yields a switching time of 50 ns. Notice that this is 250 times slower than the intrinsic rise time.

Measurement of the intrinsic-time constant can be carried out by clamping the drain-source voltage (so that no output capacitance has to be charged or discharged) and observing the output-current response to a voltage step input. By measuring the time constant of the output-current waveform, an indication of the intrinsic cutoff can be obtained. Figure 4-2 shows the circuit to be used for this measurement.

4-2 SWITCHING SPEED OF THE MOS

1. Common-source Configuration. Figure 4-3 shows the circuit of a P-channel MOS used for analyzing and measuring the turn-on time of the MOS device when the speed is limited by a capacitive load. The capacitor C represents the load. R is a large resistance used to recharge C after it has been discharged by the MOS. R is assumed to have a negligible effect upon the turn-on switching time.* Figure 4-4 illustrates the path of operation during the discharge of C, superimposed upon the MOS characteristic curves, for a step input. Assuming Q_1 is off, the capacitor charges up to voltage V_{DD}, shown as point P_1. When the gate is switched to a negative gate voltage V_G, operation moves from point P_1 to point P_2 in negligible time. From P_2 to P_3, operation is in the *saturation region and the device is assumed to be a constant current source* with a value of

$$I_D = -\frac{\beta}{2}(V_G - V_{th})^2$$

* In the design of an IC inverter, the driver device may well have significant voltage across it in the ON condition due to the static load current. The switching speed of the inverter will thus be modified by the presence of the load current. Considering I_L results in an analysis that soon becomes bogged down in algebraic manipulations. The author did not think the excessive algebra was justified, and thus the analysis considering I_L was not included. It will be shown later that the switching time in a practical inverter circuit is limited by the load device. For this situation, a more exact calculation of t_{ON} than is presented in this book is unnecessary.

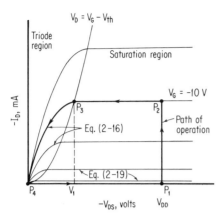

Fig. 4-4. Transient path of operation superimposed upon device characteristics.

This yields a transition time from P_2 to P_3 of

$$t = \frac{Cv}{I_D} = \frac{C|V_2 - V_1|}{|\beta|/2(V_G - V_{th})^2} \tag{4-4}$$

Equation (4-4) states the time it takes for a constant-current generator to charge a given capacitance to a given voltage. When the path of operation reaches P_3, conditions change as the device begins to operate in the triode region.

At point P_3, the model for the MOS changes to that represented by Eq. (3-19). From Fig. 4-3, the capacitor current is equated to the MOS current, resulting in

$$C\frac{dv(t)}{dt} = -\beta(V_{th} - V_{th})v(t) + \frac{\beta}{2}v(t)^2 \tag{4-5}$$

Solving this equation for $v(t)$ (the output voltage across the capacitor) gives a result that is normalized in a manner analogous to a simple R-C circuit with an exponential decay:

$$\boxed{\frac{v(t)}{V_1} = \frac{2e^{-t/\tau}}{1 + e^{-t/\tau}}} \tag{4-6}$$

where $\tau^* = C/g_m$
$g_m = -\beta(V_G - V_{th})$
$V_1 = V_G - V_{th}$

Equation (4-6) is plotted in Fig. 4-5, along with an exponential for comparison. This equation is normalized to the initial voltage V_1 (the dividing line between

---

* By judicious selection of a normalizing time constant, switching-time calculations can be greatly simplified. In the transient analysis presented in this book, equations are normalized by the time constant C/g_m. Switching times are then derived by selection of an appropriate coefficient, which may vary from 2.2 to 18. The time constant correlates the MOS circuit to a lumped, linear R-C network where R is constant at $1/g_m$ and transconductance is considered fixed at a value of $-\beta(V_G - V_{th})$. Section 3-2 dis-

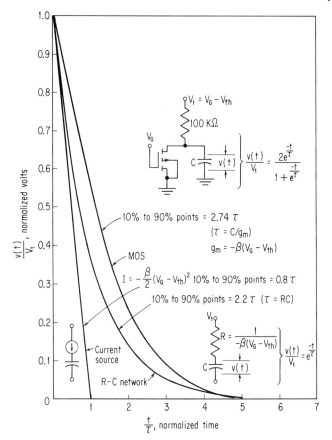

Fig. 4-5. Switching-time response of a common-source MOS driver in the triode region compared with a resistor and current-source response.

the saturation and triode regions), so that at $t = 0$, the capacitor voltage equals V_1 and the graph represents the output amplitude as unity.

Figure 4-5 compares the time response of an MOS with that of a static resistor R having a value of $1/g_m$ and with that of a current source having a value of $-\beta/2(V_G - V_{th})^2$, all discharging the same value of capacitor. The resistor R represents the initial slope (about the origin) of the V-I characteristic of the MOS device. Since the normalizing axes are the same for all three curves, they

cusses this channel resistance in some detail. It represents the initial slope (about the region $V_D = 0$) of the MOS V-I characteristic for a given gate voltage. Figure 4-6 designates the initial slope as R.

A poor choice of time constants equates the normalizing resistance to the static value illustrated in Fig. 4-6. R_1 corresponds to the supply voltage divided by the short-circuit current. Though obviously yielding the correct answer (with the appropriate coefficient), the result is considerably more unwieldy and difficult to use.

can be compared directly.   Notice that it takes the MOS a longer time to switch a given capacitor than it does the resistor R or the current source.   This is logical because the resistor R is the lowest value of channel resistance that the MOS exhibits at any time, while a current source whose value is maintained at the initial current flow is always faster in charging a capacitor.   The MOS, however, starts out with a static value that is twice the resistance of R (call it R_1).   As the drain voltage decays to zero, the static resistance looking into the drain decreases from R_1 to R.   An MOS will have a faster time response than will the resistor R_1 when both are charging the same value of capacitance.   The V-I characteristics of R, R_1, the MOS, and the applicable current source are plotted in Fig. 4-6.

The plot of Eq. (4-6) gives a result that is extremely easy to use.   Switching time is generally specified between the 10 and 90 percent points.   Starting at 0.1 and 0.9 along the normalized amplitude axis, projecting over to the curve and then downward to the normalized time axis yields a normalized time of 2.74, or in other words, $t_s = 2.74\tau$.   Thus a simple calculation of the time constant $(\tau = C/g_m)$ is all that is necessary to determine the switching time *in the triode region*.

To check the validity of Eq. (4-6), the switching time of the circuit illustrated in Fig. 4-3 was measured.   A large capacitance was used to swamp out the effects of any stray capacitance.   Operation was restricted to the triode region by setting the drain voltage at $V_G - V_{th}$, shown as voltage V_1 in Fig. 4-4.   Calculation of the switching time of this circuit will be used as an illustrative example.

A capacitor was selected which measured 0.004 μF on a capacitance bridge. Next, the necessary device parameters were found.   A graph of $\sqrt{-I_D}$ vs. V_G was

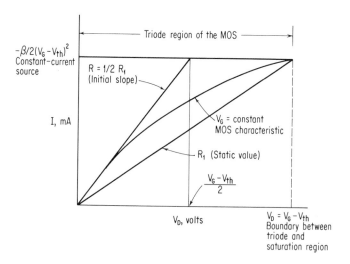

Figure 4-6

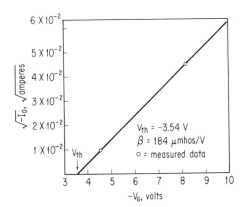

Figure 4-7

plotted from two measured values (Fig. 4-7). From this graph, V_{th} was determined. Knowing V_{th} and one measured point gave the device constant β from $I_D = -\beta/2(V_G - V_{th})^2$. The calculated value of β was 184 μmhos/V. For a -10-V gate pulse, Eq. (4-6) gives a value of

$$g_m = -(-10 + 3.54) \times 184 = 1,880 \ \mu\text{mhos}$$

The time constant of this circuit is thus

$$\frac{C}{g_m} = \frac{(0.004 \times 10^{-6})}{(1,880 \times 10^{-6})} = 2.12 \ \mu\text{s}$$

From Fig. 4-5, the switching time is predicted to be 2.74 τ, or

$$2.74 \times 2.12 \ \mu\text{s} = 5.8 \ \mu\text{s}$$

Figure 4-8a shows the input and output wave shapes. The long turn-off time of the output pulse is caused by the 100-kΩ resistor charging the 0.004-μF capacitor. A negative 10-V input pulse returns the output level to ground. The 10 to 90 percent switching-time measurement is shown in Fig. 4-8b. The measured time is approximately 6.5 μs and is in good agreement with the 5.8 μs predicted.

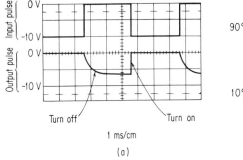

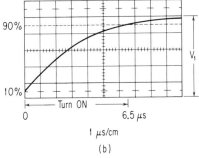

Fig. 4-8. Graphs of measured time responses for a common-source MOS switch.

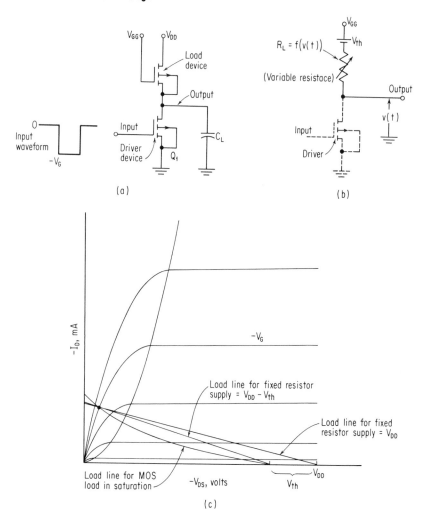

Fig. 4-9. Load switching-time circuits for saturation-region analysis: (a) inverter circuit using MOS load—substrate common to the source; (b) load equivalent circuit; (c) MOS load line and linear-resistor characteristic superimposed upon driver characteristic.

2. Load Configuration, Substrate Common to Source, Saturation Region. In ICs, it is generally desirable to use an MOS as a load resistor.* Figure 4-9 shows how a device can be biased as a load (Fig. 4-9a), gives the equivalent circuit of the device (Fig. 4-9b), and illustrates the load's nonlinear characteristic superimposed upon the output curves of the driver device (Fig. 4-9c).

The circuit used to discuss the switching time of this load arrangement is Fig. 4-9a. The driving transistor is turned on, allowing C_L to discharge completely. Q_1 is then switched off, thus effectively removing it from the circuit and allowing

* See Sec. 5-1.

the MOS load to charge C_L.* The current charging i_c is not a constant and is in fact a function of the voltage across C_L. Because the load gate is returned to the drain and the device is assumed to be an enhancement-mode device, operation is always in the saturation region. The equation for the device in this region is

$$I_D = -\frac{\beta}{2}(V_G - V_{th})^2$$

The drain and capacitor currents are equated and yield

$$C\frac{dv(t)}{dt} = \frac{\beta}{2}[V_{GG} - v(t) - V_{th}]^2$$

where $v(t)$ is the capacitor output voltage.

Solving this equation for $v(t)$ gives the time response for the MOS load and capacitor combination:

$$\boxed{\frac{v(t)}{V_1} = \frac{t/\tau}{2 + t/\tau}} \tag{4-7}$$

where $V_1 = V_{GG} - V_{th}$ (the final value)†

$\tau = C/g_m$

$g_m = -\beta(V_{GG} - V_{th})$

Even though the result is not an exponential, it is still informative to plot Eq. (4-7) in a manner similar to that shown in Fig. 4-5. Equation (4-7) is plotted in Fig. 4-10 along with an exponential response for a comparable R-C network. Notice that the MOS is considerably slower than the R-C network. This is reasonable and expected because as the output voltage across C increases toward $V_{GG} - V_{th}$, the MOS delivers less and less charging current to C. As in the resistive case [when $v(t)$ increases (decreasing the voltage across R), the current naturally decreases] the current, in the MOS case, decreases as $v(t)$ increases; however, in addition, the MOS is being turned off. This is analogous to a capacitor being charged by a nonlinear resistor whose resistance value increases as a function of capacitor voltage (see Fig. 4-9a). The speed of the circuit in Fig. 4-10, as measured between 10 and 90 percent, calculates to be 17.8τ.

An example will be presented to illustrate this method of calculating switching speed, and calculated values will then be compared with measured values. By using the same device as in the previous example, the device parameters are known from Fig. 4-7 as $\beta = 184$ μmhos, $V_{th} = -3.54$ V. Using a -10-V supply yields a g_m of $-184(-10 + 3.54) = 1,880$ μmhos and a time constant of $\tau = C/g_m = (4.1 \times 10^{-9})/(1.88 \times 10^{-3}) = 2.17$ μs.

$$\text{Switching speed} \cong 18\tau = 18 \times 2.17 \ \mu s = 39 \ \mu s$$

* Q_1 is assumed to have zero storage time.

† See footnote in Sec. 5-1.

Actual switching time was measured at 48 μs. For a -15-V supply, the g_m increases to 2,110 μmhos, which decreases the time constant to 1.94 μs. Calculated switching speed is 18×1.94 μs $= 34.9$ μs, as compared with the measured value of 40 μs. (Reference 1 covers switching speed in some detail.)

Because of the exceptionally slow response of the MOS connected as shown in Fig. 4-10, switching time of the load device, as opposed to the driver, is generally the cause of frequency limiting in digital circuits. An effective method frequently used to overcome this drawback and increase the frequency range is to return the load gate to a supply greater than $|V_{DD}|$, as illustrated in Fig. 4-11a. If V_{GG} is increased beyond the drain supply by an amount greater than the load threshold voltage, the device enters the triode region and the switching characteristic is generally improved.

3. Load Configuration, Substrate to Source, Triode Region. Increased speed can be attributed to the fact that a higher gate voltage prevents the load device from turning off during the switching transient. The higher the gate voltage, the

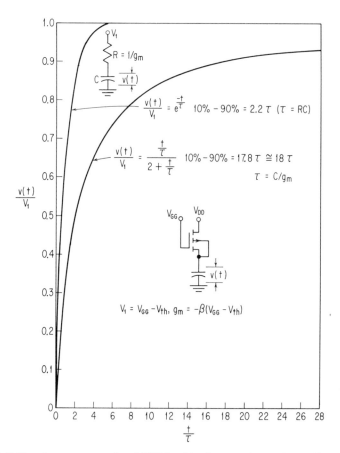

Fig. 4-10. Switching-time response of an MOS load in the saturation region, with an R-C network for comparison.

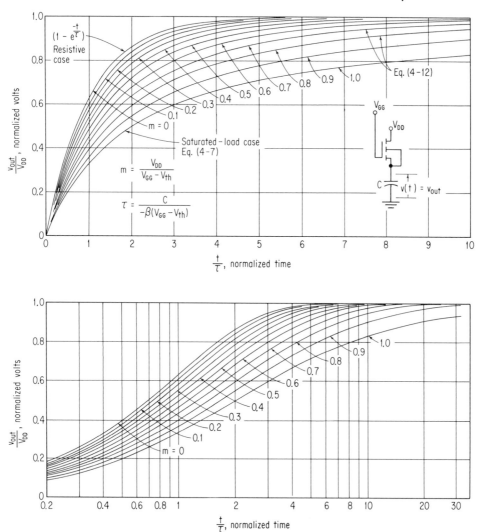

Fig. 4-11. Switching-time response of an MOS load in the triode region, showing the two limits ($m = 0$, $m = 1$) between which the MOS load must operate. When $m = 1$ the ordinate changes to $v(t)/V_1$ as in Fig. 4-10.

more linear the load V-I characteristic becomes (see Fig. 4-9c). In the limit as $V_{GG} \to \infty$, the load characteristic approaches that of a fixed resistor whose switching characteristic is 2.2τ. This represents a theoretical increase in speed of about eight times over the 18τ saturated case.

Analysis proceeds by equating the capacitor and device current (triode region). The output voltage is represented as a function of time, $v(t)$:

$$\frac{C}{\beta}\frac{dv(t)}{dt} = (V_{GG} - v(t) - V_{th})(V_{DD} - v(t)) - \frac{1}{2}(V_{DD} - v(t))^2 \qquad (4\text{-}8)$$

Rearranging the terms yields

$$\frac{C}{\beta(V_{GG} - V_{th})} \frac{dv(t)}{dt} = V_{DD} - v(t) + \frac{v(t)^2 - V_{DD}^2}{2(V_{GG} - V_{th})} \tag{4-9}$$

By dividing both sides by V_{DD}, defining $g_m = -\beta(V_{GG} - V_P)$, and letting $V_{DD}/(V_{GG} - V_{th}) \equiv m^*$, the following normalized differential equation is obtained (the m factor is a normalized bias parameter and represents how "hard" the load has been pushed into the triode region):

$$\frac{C}{g_m} \frac{d[v(t)/V_{DD}]}{dt} = 1 - \frac{v(t)}{V_{DD}} + \frac{1}{2}\left[\left(\frac{v(t)}{V_{DD}}\right)^2 - 1\right]m \tag{4-10}$$

Letting $C/g_m = \tau = C/[-\beta(V_{GG} - V_{th})]$ and separating variables yield

$$\int_0^t \frac{t}{\tau} = \int_0^{v(t)/V_{DD}} \frac{d[v(t)/V_{DD}]}{\frac{1}{2}m[v(t)/V_{DD}]^2 - [v(t)/V_{DD}] + (1 - \frac{1}{2}m)} \tag{4-11}$$

Integration of Eq. (4-11) gives the normalized output voltage as a function of the normalized time with m as a parameter:

$$\frac{v(t)}{V_{DD}} = \frac{(2 - m)(1 - e^{-(t/\tau)(1-m)})}{2 - m(1 + e^{-(t/\tau)(1-m)})} \tag{4-12}$$

Equation (4-12)† is plotted against two time scales (one linear and one log) in Fig. 4-11, where the normalized voltage parameter m results in a family of curves. The parameter m [$m = V_{DD}/(V_{GG} - V_{th})$] varies from 0 (when V_{GG} is infinitely large) to 1 (when $V_{DD} \geq V_{GG} - V_{th}$). When $m = 0$, Eq. (4-12) simplifies to the exponential that describes a fixed resistor charging a capacitor in a time of 2.2τ. When $m = 1$, Eq. (4-12) reduces (by the use of L'Hospital's rule) to Eq. (4-7), which says that the device now is operating in the saturation region with an 18τ switching time.

The information contained in Fig. 4-11 can be presented in a different form which is useful in a number of cases. The 10 to 90 percent switching time is plotted as a function of normalized voltage in Fig. 4-12. As an example, a typical case is assumed in which $V_{GG} = -24$ V, $V_{DD} = -12$ V, and $V_{th} = -5$ V. This yields an m value of 0.63. From the graph, one can see that the switching time is 4.6τ. Earlier, it was stated that switching time could be improved over the

---

* Valid for $0 \leq m \leq 1$.

† Mobility variation as given in Eq. (3-29) may be included in the switching-time measurements for more accurate results.[2] ‡ The following equation shows the added effect of $\mu(V_G)$:

$$\frac{t}{\tau_o} = \frac{1}{m - 1} \ln\left| \frac{(v(t)/V_{DD} - 1)(2 - m)}{m(v(t)/V_{DD}) - 2 + m} \right|$$
$$+ \theta(V_{GG} - V_{th}) \ln\left| \frac{(v(t))^2/V_{DD} - 2(v(t)/V_{DD}) + 2 - m}{2 - m} \right| \tag{4-12a}$$

where τ_o is the time constant using μ_o from Fig. 3-14.

‡ Superscript numbers indicate works listed in the Bibliography at the end of the chapter.

saturated case by returning the gate to a voltage higher than the drain. The example illustrates a reduction in time from 18τ to 4.6τ. Of course, the actual switching time depends upon the time constant; but for equivalent values of the time constant, C/g_m, the example shows a device switching almost four times faster in the triode region than if it were in the saturated region.

It is possible to vary the device geometry while increasing the gate voltage (for faster switching), thus keeping a constant gain (and time constant). Under this condition, the d-c power remains unchanged while the time-constant coefficient is decreased. The resultant speed increase is accomplished at no increase in the static power drain. One can also increase the gate voltage on a fixed-geometry device, resulting in an increase in gain. In this case, the time constant, as well as its coefficient, shows a decrease. Appreciably faster switching speeds are observed, but at the price of higher power dissipation.

4. Load Configuration, Substrate Common to Ground. Even though the switching curves in Fig. 4-11 were derived for an MOS load whose substrate is returned to the source terminal (discrete case), they can be used to determine "worst-case" switching speeds for the case of a grounded substrate (actual IC case). Equation (4-12) shows the switching characteristic to be dependent upon τ and m—both of which are functions of V_{th}. V_{th} is also a function of the output voltage for the grounded-substrate configuration. m and τ are shown in the following equation to be functions of the variable pinchoff voltage:

$$m = \frac{V_{DD}}{V_{GG} - (V_{th} + \Delta V_{th})}$$

$$\tau = \frac{C}{-\beta[V_{GG} - (V_{th} + \Delta V_{th})]} \tag{4-13}$$

(Figure 2-16 plots ΔV_{th} as a function of back-gate bias or output voltage.)

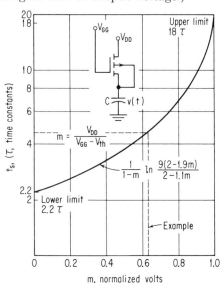

Fig. 4-12. Switching time (10 to 90 percent points) vs. biasing parameter m.

Notice in Eq. (4-13) that as $|\Delta V_{th}|$ increases, the terms m and τ also increase. Figure 4-12 shows that an increase in these two terms results in increased switching times. A "worst-case" approach to grounded-substrate switching-time calculations would be:

1. Consider the highest output voltage swing.
2. Using this voltage, refer to Fig. 2-16 to find the value of ΔV_{th}.
3. Place this voltage into Eq. (4-13) to find the "worst-case" values of m and τ.
4. Read the switching time from Fig. 4-11a or 4-12.

To illustrate the above procedure and to check the validity of the switching-time calculations, measured times will be compared with calculated times for both the grounded-substrate case and the case of the substrate returned to the source. Measurements on device No. 78B-7 gave the following parameters:

1. $V_{th} = 5.4$ V (extrapolated value).
2. $\beta = 50$ μmhos/V.
3. $K_1 = 1.1$ V$^{1/2}$.

The device was operated in the triode region with a drain voltage of -10 V for two values of gate voltage: -25.4 V and -35.4 V. A capacitor measuring 786 pF was used as the load. Stray capacitance was assumed equal to 9 pF, giving a total capacitive load of 795 pF for calculations. Actual switching time without the external capacitor was very fast (50 ns) in comparison with any of the actual switching times measured, so that stray effects were neglected.

Time-constant calculations require the C/β ratio, which was found to be

$$\frac{C}{-\beta} = \frac{795 \times 10^{-12}}{-50 \times 10^{-6}} = -15.9 \ \mu s - V \tag{4-14}$$

For the first measurement (grounded substrate, $V_{GG} = -24.5$ V):

$$m = \frac{-10}{-25.4 + 5.4} = 0.5 \qquad \tau = \frac{-15.9 \ \mu s - V}{(-25.4 + 5.4) \ V} = 0.795 \ \mu s$$

Figure 4-12 shows the 10 to 90 percent rise time to be 3.75τ. Thus

$$t_r = 3.75(0.79) \ \mu s = 2.98 \ \mu s$$

The actual measured value was 2.5 μs. For the second measurement, the gate voltage was increased to -35.4 V, yielding

$$m = \frac{-10}{-35.4 + 5.4} = 0.33 \qquad \tau = \frac{-15.9 \ \mu s - V}{(-35.4 + 5.4) \ V} = 0.53 \ \mu s$$

From Fig. 4-12, the rise time is equal to 3.0τ:

$$t_r = 3(0.53) \ \mu s = 1.59 \ \mu s$$

The measured value was 1.55 μs.

The third case considers the grounded-substrate condition. Referring to Fig. 2-16, for an output swing of -10 V and $K_1 = 1.1$, $\Delta V_{th} = -2.6$ V. Threshold voltage is to be increased by -2.6 V, so that $V_{th}(v_{out}) = -5.4 - 2.6 = 8.0$ V. Again, for $V_{GG} = -24.5$,

$$ m = \frac{-10}{-24.5 + 8} = 0.575 \qquad \tau = \frac{-15.9 \ \mu s - V}{(-24.5 + 8) \ V} = 0.915 \ \mu s $$

From Fig. 4-12, $t_r = 4.2\tau$. Calculated rise time is then

$$ t_r = 4.2(0.915) \ \mu s = 3.84 \ \mu s $$

Rise time was measured at 3.3 μs. Also, for the grounded-substrate case, a fourth measurement is made where $V_{GG} = -35.4$ V:

$$ m = \frac{-10}{-35.4 + 8} = 0.365 \qquad \tau = \frac{-15.9 \ \mu s - V}{(-35.4 + 8) \ V} = 0.58 \ \mu s $$

Rise time is 3.12τ. $t_t = 3.12(0.58) \ \mu s = 1.82 \ \mu s$. This value compares favorably with the 1.9 μs measured time.

Conclusions regarding the load switching times are as follows:

1. Equations and curves given in this chapter form a reasonably accurate model for switching considerations.
2. The argument presented, which takes into account the grounded substrate, is a valid argument yielding good results.
3. The lower the gate voltage, the less accurate are the switching times. This is due to the difficulty in describing threshold voltages at low current levels. (See the footnote regarding threshold voltage in Sec. 5-1.) Threshold-voltage variations also show up as a smaller percentage of error at high gate voltages than at lower voltages.
4. The following table summarizes the calculated and measured switching times. The percentages shown represent the error in the calculated values as compared with the measured values.

	$V_{GG} = -25.4$	$V_{GG} = -35.4$
Substrate to source:		
Calculated.............	2.98 μs	1.59 μs
Measured.............	2.5 μs	1.55 μs
Error................	19.2%	2.6%
Substrate to ground:		
Calculated.............	3.84 μs	1.82 μs
Measured.............	3.3 μs	1.9 μs
Error................	15.5%	4.2%

The photographs in Fig. 4-13 show composite switching times for the MOS load. Four values of gate voltage are used to illustrate switching-time dependence upon this voltage. Notice the deleterious effect the back-gate bias (output volt-

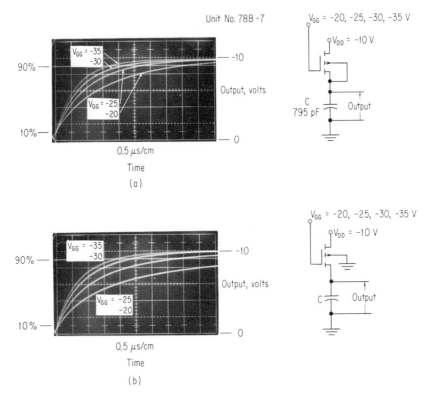

Fig. 4-13. Photographs of actual MOS load switching times: (*a*) substrate common to source; (*b*) substrate common to ground.

age) produces in switching times (shown in Fig. 4-13*b*). This is particularly evident at the lower gate voltages, where ΔV_{th} causes a larger percentage of change in m and τ than at the higher voltages.

BIBLIOGRAPHY

1. Burns, J. R.: Switching Response of Complementary-symmetry MOS Transistor Logic Circuits, *RCA Rev.*, pp. 627–661, December, 1964.
2. Personal communication with R. P. Thiels.   Texas Instruments Incorporated.

GENERAL REFERENCE

O'Reilly, T. J.: The Transient Response of Insulated-gate Field-effect Transistors, *Solid-state Electron.*, vol. 8, pp. 947–956, December, 1965.

5

Basic MOS Integrated-circuit Concepts

Until now, this book has discussed the MOS as if it were essentially a discrete device. The device has been characterized, equations have been written, and various factors—those that dominate device operation—have been considered. However, as subtitle suggests, the MOS must also be considered as an IC circuit element. In this age of ICs, the major impact of MOS technology will be made in this area, and it is fitting that this chapter should emphasize the relationship of this new technology to IC usage.

Upon examining the physical layout and structure used in MOS integrated technology, one immediately notices the total lack of conventional components in integrated form. For the most part, large MOS integrated arrays contain only MOS devices—no individual resistors, capacitors, or diodes.* There are a number of reasons for this departure from the use of conventional components:

1. The MOS can effectively perform the function of a diffused resistor.
2. Coupling capacitors are omitted because of the ability of the MOS to direct-couple.
3. Multiple-clocking schemes, which are used in the great majority of MOS circuits, help eliminate components such as blocking capacitors.
4. The symmetrical nature of the MOS allows it to be used as a symmetrical switch—giving the designer an added degree of freedom not generally available in bipolar ICs.

5-1 THE MOS AS A LOAD RESISTOR†

While diffused resistors typically exhibit 100 to 200 $\Omega/\square$, the MOS is capable of 20,000 $\Omega/\square$, giving practical resistance values of the order of 100 to 200 kΩ. By

---

* The exceptions are protective diodes across the input gates and stray capacitance used to store charge in "dynamic" systems.

† For an additional discussion of the MOS as a load, see Ref. 19 in the Bibliography for Chap. 2.

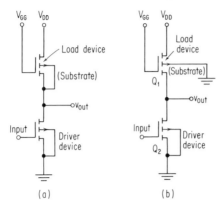

Fig. 5-1. Inverters using MOS loads: (a) discrete case; (b) IC case.

using less area than a conventional diffused resistor, MOS technology allows more complex circuitry on a single monolithic chip than would otherwise be possible.

1. Load Equivalent Circuits.   Figure 5-1 shows an inverter circuit utilizing an MOS as the load resistor.   Two possibilities for connecting the substrate are shown.   In Fig. 5-1a, the substrate is connected to the source (this is possible when discrete devices are used); while in Fig. 5-1b, the substrate is grounded (which is always the case for MOS devices in integrated form).

For enhancement-mode devices and for $V_{GS} = V_{DS}$, the load is restricted to the saturation region.   The inequality determining saturation-region operation demands that $|V_D| \geq |V_G - V_{th}|$, and when $V_G = V_D$, operation is always to the right of the $V_D = V_G - V_{th}$ curve (shown as the dashed line in Fig. 5-2).   The equation describing operation in this region is

$$I_D = -\frac{\beta}{2}(V_G - V_{th})^2 \tag{5-1}$$

The V-I load line characteristic of a P-channel device (Fig. 5-1) superimposed

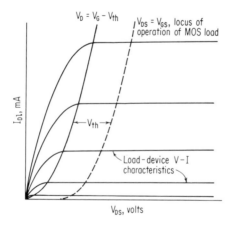

Fig. 5-2. Load-device characteristics showing locus of operation in the saturation region.

upon the characteristic curve of a driver device is shown in Fig. 5-3. This figure, of course, represents the graphical solution of two simultaneous equations; one is the equation for the driver device, and the other is the equation for the load device. Any intersection of these two curves represents a possible operating condition.

In the ON condition, operation is shown at point P_1. Low-current operation and a large-geometry driver device are assumed. The voltage across the driver will be small when compared with the supply, so that $V_{ON} \cong 0$ V. Also I_{ON}, the current actually flowing, is assumed equal to I_1, which is the current that would actually flow if $V_{ON} = 0$. When the driver is switched off, the output voltage (Fig. 5-1) rises toward the gate voltage minus the threshold voltage of the load ($V_{out} \cong V_{DD} - V_{th}$, when $V_{GG} = V_{DD}$). Notice that the output-voltage swing is less than the full supply voltage. For example, if $V_{th} = -5$ V, $V_{DD} = -15$ V, the output of the inverter (Fig. 5-1a) only swings from 0 to -10 V. The reduction in output occurs because a voltage approximately equal to V_{th} must be dropped across the load (gate-to-source) to initiate conduction. To solve for the static operating conditions of the circuits in Fig. 5-1, equivalent circuits will be derived for both the ON and OFF conditions.

If the driver device of Fig. 5-1 is on hard enough to be assumed a short circuit ($V_{ON} \cong 0$ V), then the current due to the load device is given in Eq. (5-1). Since the gate is returned back to the drain supply, $V_G = V_{DD}$, and Eq. (5-1) becomes

$$ I = -\frac{\beta}{2}(V_{DD} - V_{th})^2 \tag{5-2} $$

This equation can be rearranged to give a physical interpretation of the d-c load in terms of a Thévenin's equivalent circuit:

$$ I = \frac{V_{DD} - V_{th}}{2/[-\beta(V_{DD} - V_{th})]} = \frac{V_{DD} - V_{th}}{2/g_{ml}} = \frac{\Delta V}{R_L} = \frac{V_{DD} - V_{th}}{R_L} \tag{5-3} $$

Equation (5-3) says that the load current may be thought of as a voltage divided

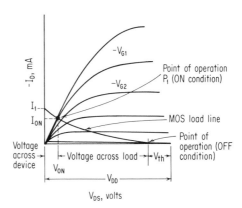

Fig. 5-3. Driver characteristics with MOS load line superimposed.

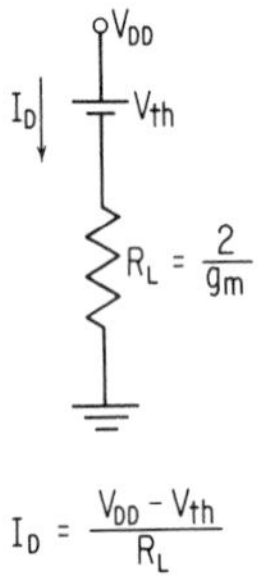

$$I_D = \frac{V_{DD} - V_{th}}{R_L}$$

Fig. 5-4. Equivalent ON circuit of MOS load in the saturation region.

by a given resistance; the voltage is the drain supply reduced by the threshold voltage, and the load resistor is $2/g_{ml}$. (The g_m here is, of course, that of the load.) The equivalent circuit given by Eq. (5-3) is shown in Fig. 5-4. For example, assume a -15-V drain supply, a -6-V threshold, and a typical g_m (for a P-channel IC load) of 15 μmhos. The static or d-c current is*

$$I = \frac{-15 + 6}{+2/15 \times 10^{-6}} = \frac{-9 \text{ V}}{133 \text{ k}\Omega} = -67.5 \ \mu\text{A}$$

From the above discussion,

$$\boxed{R_L = \frac{2}{g_{ml}}} \tag{5-4}$$

A graphical interpretation of Eq. (5-3) and Fig. 5-4 is seen by observing the *V-I* characteristics of a typical load device, as shown in Fig. 5-5. In the equivalent circuit, a voltage V_{th} is shown opposing the supply—this is represented by displacing the *V-I* curve from the origin by an amount V_{th}. The resistance of the equivalent circuit is the reciprocal slope of the line from P_1 (the new origin for the curve) to P_2, the d-c operating point. Because a static (as opposed to incremental) resistance is being considered, the reciprocal slope of the line P_1-P_2 is chosen as the equivalent resistance instead of the tangent at point P_2. Since the load is operating in the saturation region, the *V-I* characteristic is a square-law or parabolic curve. Due to the mathematics of this type of curve, $R_L = 2/g_{ml}$, where g_{ml} is the slope at P_1.

The equivalent circuit of the active load in the OFF condition is not so easily or "nicely" characterized as it is in the ON state. However, it is still reasonable to represent the load as some resistor in series with some voltage. When the driver device of Fig. 5-1 turns off, the output voltage (the load source) begins to rise.

---

* V_{ON} is the product of the load current and the resistance of the driver device. This can be expressed as $(V_{DD} - V_{th})g_{ml}/(g_{ml} + 2g_{mD})$. When $R_{load} \gg R_{driver}$, V_{ON} can be written as

$$V_{ON} \cong \tfrac{1}{2} \frac{g_{ml}}{g_{mD}} (V_{DD} - V_{th}) \tag{5-5}$$

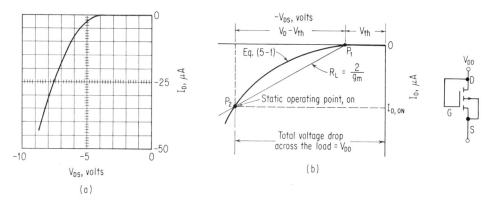

Fig. 5-5. $V-I$ characteristics of a two-terminal MOS load (saturation region).

This reduces the gate-to-source voltage across the load and begins turning the device off. The load transconductance is reduced, which in turn increases its resistance. This resistance increases until, theoretically, it reaches an infinite value when the load has turned itself off. Of course, there are practical bounds on the output resistance, such as the resistance of the source-to-substrate reverse-biased P-N junction.

2. Output Voltage of a Load Device. Figure 5-1 shows the two cases that must be considered when discussing the d-c output voltage (the final value) of the load. In the discrete case, where the substrate can be returned directly to the source, the pinchoff or threshold voltage is not modulated by a reverse bias on the channel (see Sec. 2-3). For this case, the output voltage is simply $V_{GG} - V_{th}$ (where V_{th} is constant).* However, in the IC case, the load and driver substrate are common and at ground potential. As the source terminal of the load (the output) rises, the channel becomes reverse-biased with respect to the substrate and the threshold varies in accordance with Eq. (2-31). The output swing is thus reduced in magnitude from the situation shown in Fig. 5-1a. The output is now $V_{GG} - V_{th}(v_{out})$,* where threshold is shown to be a function of the output voltage.

By using the approximation that $v_{out} = V_{GG} - V_{th}(v_{out})$ and substituting Eq.

---

* These two equations representing the output voltage are of necessity only approximations. In the low-current region, threshold or pinchoff is difficult to define and express analytically. Threshold is generally considered, in this book, to be an extrapolated value, so that the $V-I$ characteristic of the device does not actually pass through V_{th}. The extrapolated line that defines threshold is determined from data points in a relatively high-current region. As the actual device current approaches threshold, it departs from the straight-line representation (see Figs. 2-14 and 2-16). One can see that in the low-current region (where the OFF load operates), the amount of current will modify the output voltage found by using the extrapolated V_{th}. Thus, in Fig. 2-16, the magnitude of the threshold voltage at -0.5 μA is 0.25 V lower than the extrapolated value. In an IC, the only significant d-c load being driven by internal active MOS loads will be the leakage current of reverse-biased junctions. This type of load is difficult to predict in a way that will lend itself to easy circuit design. The author has found, however, that reasonable results can be expected when the extrapolated value of pinchoff is used.

(2-30) for $V_{th}(v_{out})$, it is possible to derive a simple expression relating the input (V_{DD}) and output (v_{out}) voltages as a function of doping level, oxide thickness, etc. (constant K_1):

$$v_{out} = V_{GG} - (V_{th} + \Delta V_{th}) \tag{5-6}$$

Rearranging terms and solving for $V_{GG} - V_{th}$ as a function of the output voltage yield

$$V_{GG} - V_{th} = v_{out} + \Delta V_{th}$$

or

$$V_{GG} - V_{th} = v_{out} - K_1[\sqrt{-(2\phi_F + v_{out})} - \sqrt{-2\phi_F}] \tag{5-7}$$

Equation (5-7) is left in terms of the output voltage strictly for convenience. Solving for v_{out} results in a large and unwieldy expression which is more difficult to plot than Eq. (5-7).

As an example, assume an output of -13 V, a threshold of -5 V, and $K_1 = 1.0$. From Fig. 2-16, $\Delta V_{th} = -2.9$ V. Plugging these values into Eq. (5-7) yields

$$V_{GG} + 5 = -13 - 2.9$$
$$V_{GG} = -20.9 \text{ volts}$$

Equation (5-7), together with the datum point from the example, is plotted in Fig. 5-6a for the case where $V_{GG} = V_{DD}$. v_{out} is plotted against both ($V_{GG} - V_{th}$), a normalized abscissa which makes the graph a universal graph, and V_{GG} for a specific illustration where $V_{th} = -5$ V. Notice that as the substrate doping is increased, higher input voltages V_{GG} are required to produce the same output voltage. This figure also shows that the output voltage is independent of the drain voltage so long as the drain supply is kept high.

3. Load Gate Returned to Separate Supply. Figure 5-6a shows that a significant percentage of the total supply voltage can be lost across an active MOS load. For the example illustrated in the figure, the output voltage is reduced 38 percent below the input. This situation results in unnecessary power dissipation to achieve a desired output voltage. Fortunately, there is an easy solution to the problem. By returning the gate supply to a higher voltage than V_{DD}, threshold voltage is effectively cancelled. When the gate supply V_{GG} is returned high enough to offset the deleterious effect of $V_{th} + \Delta V_{th}$, the load source (the output) will swing the full drain supply and clamp at V_{DD}, as shown in Fig. 5-6b. Notice that the abscissa is plotted as $V_{GG} - V_{th}$ to normalize the curve to the general case.

Verification of Eq. (5-7) is seen in Fig. 5-6c. Here the device parameters K_1 and V_{th} were measured and plugged into the equation, and v_{out} was plotted. Measured data of the output voltage were also plotted. Close agreement is observed between the calculated and measured voltages. This suggests that using extrapolated values of threshold in Eq. (5-7) does, in fact, yield reasonably close results.

5-2 MOS INVERTER

1. Physical Layout. The inverter circuit of Fig. 5-1b is shown implemented in a physical layout in Fig. 5-7. The driver device Q_2 is shown as a large-geometry

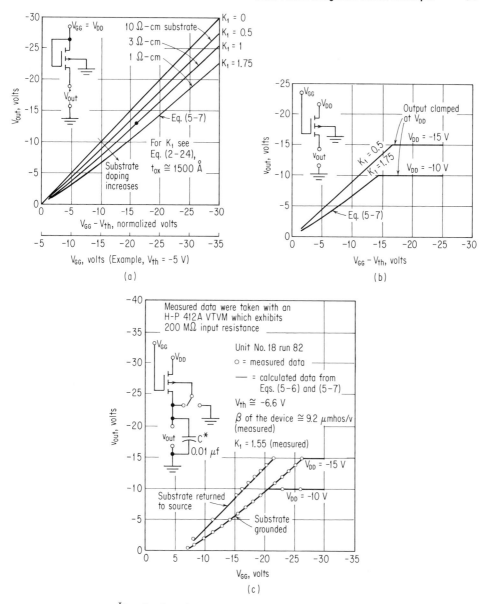

*Capacitor C was included to reduce noise during the measurement

Fig. 5-6. Output voltage from MOS load with grounded substrate. (*a*) **Gate returned to drain.** (*b*) **Gate returned to a supply that is higher than the drain.** (*c*) **Measured data compared with theory.**

device when compared with the load device Q_1. The bottommost P$^+$ diffusion (heavy P diffusion) serves as the source of the driver device. The top P$^+$ diffusion serves as the drain of the load, while the middle diffusion acts as both a source and a drain for the load and driver, respectively. This utilization of a common diffused area for more than one function, serving as both a source and a drain, greatly reduces the overall chip area when fabricating MOS circuits. Another space saving in MOS circuitry results because no isolation diffusions are required between devices; the P$^+$ regions are inherently isolated from the N-type substrate.

Fringing effects between the ends of the source and drain present a design problem since there exists a portion of the channel that is uncontrolled by the device gate. This problem can be solved by the use of two techniques: (1) terminating the channel ends in thick oxide (12,000 to 15,000 Å), and (2) extending the gate metal past the device diffusions on each end, as shown in Fig. 5-7.

The effect of using thick oxide is seen in Eq. (2-24). This equation shows that threshold voltage is directly proportional to oxide thickness t_{ox}. Thus the thicker the oxide, the higher the turn-on voltage. Assume the driver device exhibits a 4-V threshold and uses 1,500 Å of gate oxide. If the ends of the channel are terminated in 15,000 Å oxide, these portions of the wafer (external to the device) will show a threshold 10 times greater than the driver, or 40 V. Limiting voltages to less than 40 V ensures that no silicon-surface inversion will take place in thick-oxide regions. No inversion means no fringing channel.

The reason for extending the gate metal past the ends of the channel is to be sure to maintain control over the fringing channel, if it does, in fact, exist. The metal overlaps sufficiently that any fringing channel that does exist will have such a low W/L ratio that its effect will be extremely small.

In the layout of Fig. 5-7, the load gate is returned directly to the load drain and

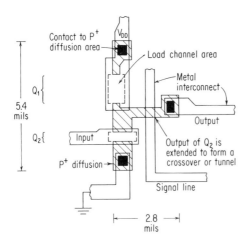

Fig. 5-7. Physical layout of practical inverter (scale drawing).

V_{DD}.   This has the advantage of requiring only one power supply and one inter-connecting lead.   There are times in MOS integrated circuitry that any dis-advantage the load may exhibit when wired as shown in Fig. 5-7 is outweighed by the topological problem presented when just one additional bus wire must be incorporated into the layout.

2. Design Example.   It will now be constructive to design a simple inverter circuit.   The configuration is to be the same as in Fig. 5-1b, and the inverter is to be built on 3-Ω-cm N-type silicon ($K_1 \cong 1$).   A voltage swing of from -0.5 to -10 V is required, and the ON current is to equal -50 µA.   Threshold voltages are typically -5 V.   The input-voltage swing is from -0.5 to -10 V.

   a. Driver-device Calculation.

$$V_{ON} = I_D r_{ON}$$

$$\therefore \qquad r_{ON} = \frac{V_{ON}}{I_D}$$

$$r_{ON} = \frac{-0.5 \text{ V}}{-50 \text{ µA}} = 10 \text{ k}\Omega$$

where r_{ON} is the ON resistance of the driver.

   From Eq. (3-21), $1/r_{ON} \cong g_m$, and

$$g_m = \frac{1}{10 \text{ k}\Omega} = 100 \text{ µmhos}$$

From Eq. (3-3), g_m is given as

$$g_m = -\beta(V_G - V_{th})$$

or

$$g_m = \frac{\mu_p \epsilon_{ox}}{t_{ox}} \frac{W}{L} (V_G - V_{th})$$

Using typical values for the constants (see the Notation section in the beginning of the book) gives a numerical value of

$$g_m = -4.5 \times 10^{-6} \frac{W}{L} (-5 \text{ V})$$

Assuming a channel length of 0.2 mil yields a channel width of 0.89 mil for a width-to-length ratio of 4.45.

   b. Load-device Calculation (Saturation Region).   Referring to Fig. 5-6 for the output voltage, a -17.5-V supply is required for -10 V out. $I_{D(ON)}$ can be calculated as follows:

   From Eq. (5-3), $I_{D(ON)} = (V_D - V_{ON} - V_{th})/R_L$, where $V_D = -17.5$ V, $V_{th} = -5$ V, $V_{ON} = -0.5$ V.

$$R_L = \frac{-17.5 + 0.5 + 5}{-50 \text{ µA}} = 0.24 \text{ M}\Omega$$

Now, from Eq. (5-4), $R_L = 2/g_{ml}$, so that

$$g_{ml} = \frac{2}{R_L} = \frac{2}{0.24 \times 10^6} = 8.34 \ \mu\text{mhos}$$

$$g_m = -4.5 \times 10^{-6} \frac{W}{L} (-17.5 + 5 + 0.5) = 8.34 \times 10^{-6}$$

Assuming a width of 0.2 mil yields a length of 1.3 mils for a W/L ratio of 0.154.

Notice the vast difference in the character of the geometries between the two types of devices. The W/L ratio of the driver device is much greater than that for the load. In fact $(W/L)_D/(W/L)_l = 28.9$. As will be demonstrated shortly, the greater the ratio of $(W/L)_D/(W/L)_l$, the greater will be the voltage gain of the inverter and, consequently, the noise margin.

3. Static Transfer Curves, Substrate Returned to Source. A good deal of information concerning the operation of the inverter stage of Fig. 5-1a is contained in its static transfer curve, where the output voltage is plotted as a function of the input voltage. Analysis is based upon Fig. 5-8, in which the nonlinear load line of the MOS load is superimposed upon the V-I characteristic of the driver. In this low-frequency analysis, capacitive reactances are considered negligible, so the load and driver currents may be equated:

$$-\frac{\beta_l}{2}[(V_{DD} - v_{DSD}) - V_{thl}]^2 = -\frac{\beta_D}{2}(V_{GSD} - V_{thD})^2 \qquad (5\text{-}8)$$

where $V_{GG} = V_{DD}$. (The left-hand side corresponds to the load current, while the right-hand side represents the driver.) From Fig. 5-8, it is seen that V_{DSD} is

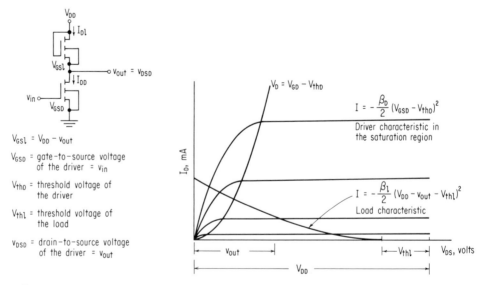

Fig. 5-8. Load line characteristic of active load superimposed upon driver-device characteristic.

the output voltage, while V_{GSD} corresponds to the inverter input voltage.   Substituting v_{out} and v_{in} into Eq. (5-8) and solving for the normalized output voltage as a function of the normalized input voltage yield

$$\frac{v_{out}}{V_{GG} - V_{thl}} = -\sqrt{\beta_R}\,\frac{v_{in} - V_{thD}}{V_{GG} - V_{thl}} + 1 \tag{5-9}$$

where $\sqrt{\beta_R} = \sqrt{\beta_D/\beta_l}$.

Notice that this is the equation for a straight line.   Equation (5-9) is valid so long as the driver device stays in the saturation region.   When the driver enters the triode region, its current equation changes to $I_D = -\beta_D[(v_{in} - V_{thD})v_{out} - v_{out}^2]$. Again equating the driver and load currents results in a normalized expression for the transfer curve:

$$\frac{v_{out}}{V_{GG} - V_{thl}} = \frac{1 + \beta_R\dfrac{v_{in} - V_{thD}}{V_{GG} - V_{thl}} \pm \sqrt{\left(\beta_R\dfrac{v_{in} - V_{thD}}{V_{GG} - V_{th}}\right)^2 + \beta_R\left(2\dfrac{v_{in} - V_{thD}}{V_{GG} - V_{th}} - 1\right)}}{1 + \beta_R} \tag{5-10}$$

where $\beta_R = \beta_D/\beta_l$; a minus sign is used for a P-channel device.

Equations (5-9) and (5-10) are presented in normalized form in Fig. 5-9.   For a practical example, assume $V_{thl} = V_{thD} = -5$ V and $V_{DD} = V_{GG} = -15$ V.

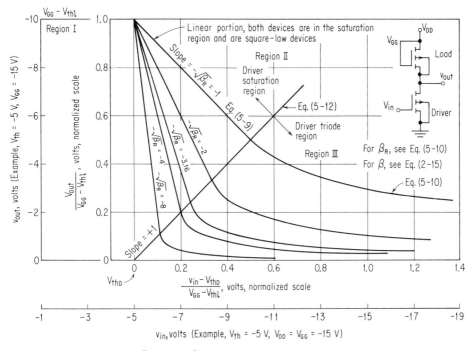

Fig. 5-9. Static transfer curves; v_{out} vs. v_{in}.

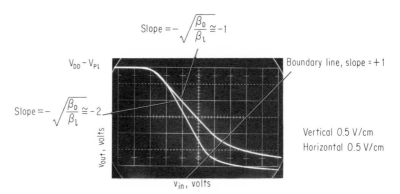

Fig. 5-10. Photograph of superimposed transfer curves.

The output remains constant at -10 V as the magnitude of v_{in} is increased from 0 to -5 V. At this point, the driver begins to conduct. Both the driver and the load are in the saturation region (region II), where the output voltage is a linear function of the input voltage. The slope of the curves in region II is found by differentiating Eq. (5-9) with respect to v_{in}. The resulting equation, an expression for the voltage gain in the linear transition region, is dependent upon device geometries only and not on the applied voltage:*

$$\frac{dv_{\text{out}}}{dv_{\text{in}}} = A_v = -\sqrt{\frac{\beta_D}{\beta_l}} = -\sqrt{\frac{(W/L)_D}{(W/L)_l}} \tag{5-11}$$

The physical reason for the straight-line relationship in region II is that both devices are square-law devices operating in the saturation region. The square-law characteristics cancel in such a way as to produce a linear relationship.

As $|v_{in}|$ increases, operation shifts into region III, where the driver device begins to operate in the triode region. Here there is no longer a linear correspondence between v_{out} and v_{in}. Regions II and III are separated by a line originating from the driver threshold point (in the case of the example, -5 V). The equation for this dividing line is

$$v_{\text{out}} = v_{\text{in}} - V_{PD} \tag{5-12}$$

which is essentially the same as Eq. (2-18).

Figure 5-10 shows two transfer curves superimposed on the same photograph. The devices used for these measurements had identical geometries. The uppermost curve is for the case of one active load and one identical driver. Because of equal width-to-length ratios, the |gain| (or |slope|) equals unity ($\sqrt{1/1} = 1$). The lower curve represents the situation of one load common to four driver devices in parallel. With four times the width for the driver device, the gain or slope increases to 2 [see Eq. (5-11)]. The measured slope of the transfer curves agrees

---

* Equation (5-11) suggests a method for determining the W/L ratio of a device when device constants, such as t_{ox} [see Eq. (2-15)], are unknown.

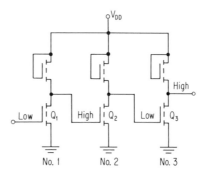

Fig. 5-11. Circuit used to define noise margin.

closely with the predicted values, and the general shape of the curves follows closely those of Fig. 5-9.*

The foregoing analysis tacitly assumes that the threshold voltage of the load device remains constant as the output voltage swings with respect to ground or the substrate. This is not strictly true in the case of ICs (Fig. 5-1b), but it has been used to introduce the idea of the static transfer curve.

The curves of Fig. 5-9 are useful in predicting the noise margin of the inverter circuit.† Noise margin may be defined loosely as the ability of a circuit to give the correct output in the presence of spurious signals. Figure 5-11, showing three simple inverters connected in cascade, will be used to illustrate the concept of noise margin. Because the drivers are assumed to be identical and the loads are assumed to be equal, one transfer curve (Fig. 5-12) will serve for all three inverters. As a practical matter, in the following approximate analysis, noise margin is defined as the voltage necessary to move the point of operation into the high-gain transition region (region II).

The noise margin of inverter No. 3, the OFF inverter, is simply the difference

---

* It is obvious that the circuit shown in Fig. 5-9 would be useful in amplifier design, where a linear output as a function of input voltage is required. A large linear swing of the output can be achieved in a simple inverter circuit without the use of negative feedback. See Sec. 6-4 for a further discussion of the linear application.

† For an additional discussion of noise margin, see Ref. 1 in the Bibliography at the end of the chapter.

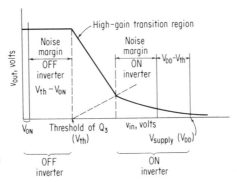

Fig. 5-12. Inverter transfer curve illustrating noise margin.

between the threshold voltage of Q_3 and the ON voltage of Q_2. In other words, the noise margin is that voltage which, when added to the normal voltage on the gate of Q_3 (which is V_{ON} of Q_2), will initiate conduction in Q_3. At this point, operation enters the transition region. Noise-margin voltage is thus $V_{th} - V_{ON}$ and is shown graphically in Fig. 5-12.

Now, considering the case of the ON inverter (No. 2), one can see that the supply minus the load threshold voltage of inverter No. 1 is the input voltage driving the ON inverter. The difference between this driving voltage and the voltage at the edge of the high-gain transition region is the noise margin of the ON inverter.

It should be pointed out that the definition of noise margin used (i.e., the voltage required to bring the operating point just to the edge of the transition region) is a conservative criterion in practical circuits. In most cases, spurious signals can, in fact, push the operating point somewhat into the transition region without producing a false output. As the gain of the driver device is increased, the slope of the transition region in Fig. 5-12 increases, so that the approximate analysis approaches the actual noise margin.

The main point that can be seen from Fig. 5-12 is that noise margin increases as the gain of the driver is increased with respect to the load. Therefore, to design a high-noise-immunity circuit, keep the driver-load W/L ratio high.

4. Static Transfer Curves, Substrate Returned to Ground. In IC design, both load and driver devices are fabricated on a common piece of silicon, so that the substrates of both devices are common and are at ground potential. This leads to a modulation of the pinchoff voltage as the output swings. Calculation of the transfer curves follows the previous analysis, i.e., equating the driver and load currents. In addition, threshold is considered to be a function of the output voltage [see Eq. (2-30)]. For the case of an inverter stage with a saturated load, the following expressions are applicable:

$$v_{in} - V_{thD} = \frac{1}{\sqrt{\beta_R}} [(V_{GG} - V_{thl}) + K_1(\sqrt{-(2\phi_F + v_{out})} - \sqrt{-2\phi_F}) - v_{out}]$$

$$\text{or } v_{in} - V_{thD} = \frac{1}{\sqrt{\beta_R}} [(V_{GG} - V_{thl}) - \Delta V_{th} - v_{out}] \tag{5-13}$$

(valid for region II, that is, $|v_{out}| \geq |v_{in} - V_{thD}|$); and

$$v_{in} - V_{thD} = \tfrac{1}{2} v_{out} + \frac{\tfrac{1}{2}[(V_{GG} - V_{thl}) - v_{out} + K_1(\sqrt{-(2\phi_F + v_{out})} - \sqrt{-2\phi_F})]^2}{\beta_R v_{out}}$$

$$\text{or } v_{in} - V_{thD} = \tfrac{1}{2} v_{out} + \frac{\tfrac{1}{2}[(V_{GG} - V_{thl}) - v_{out} - \Delta V_{th}]^2}{\beta_R v_{out}} \tag{5-14}$$

(valid for region III, that is, $|v_{out}| < |v_{in} - V_{thD}|$).

These equations give the input voltage as a function of the output voltage. Normally, one would solve for the output as a function of the input. However,

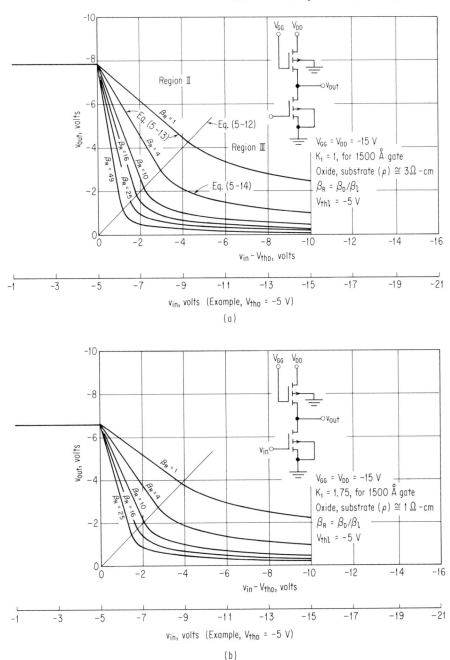

Fig. 5-13.   Transfer curves for saturated loads with substrates common to ground.

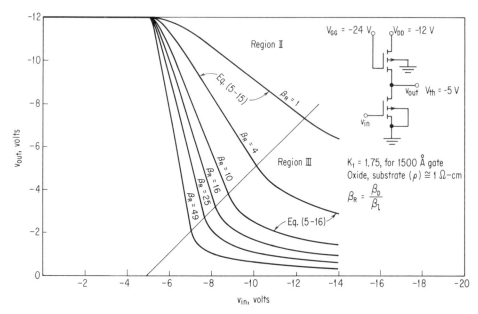

Fig. 5-14. Transfer curve for triode-region load operation with substrate returned to ground.

the amount of algebra necessary to transform the above equations is excessive, and any additional information gained does not justify (for the author) changing the simple results of Eqs. (5-13) and (5-14). To use these results, one simply selects a value of output voltage, finds ΔV_{th} from Fig. 2-16, and solves for the input voltage. These equations are plotted in Fig. 5-13 for two values of starting substrate resistivity.

In comparing the curves of Fig. 5-13 with those of Fig. 5-9, three main differences are evident:

1. The maximum output voltage is greater in the case of the substrate common to the source terminal.
2. The slope (or gain) of the transfer curve in region II is greater in Fig. 5-9 (for a given β_R) than in the case where the substrate is grounded.
3. For a given input voltage and β_R ratio, the output voltage (V_{ON}) is lower in Fig. 5-13 than in Fig. 5-9. This is because an increase in the threshold voltage reduces the load current, resulting in a lower IR drop across the driver device.

The same type of analysis can be carried out on an inverter stage with its load biased in the triode region, as shown in Fig. 5-14. The following equations describe a family of transfer curves:

$$v_{\text{in}} - V_{thD} = -\frac{1}{\sqrt{\beta_R}} \sqrt{2(V_{DD} - v_{\text{out}})}$$

$$\times \sqrt{V_{GG} - V_{thl} + K_1[\sqrt{-(2\phi_S + v_{\text{out}})} - \sqrt{2\phi_F}] - \frac{V_{DD^2} - v_{\text{out}}^2}{2(V_{DD} - v_{\text{out}})}} \quad (5\text{-}15)$$

$$\text{or } v_{\text{in}} - V_{thD} = -\frac{1}{\sqrt{\beta_R}} \sqrt{2(V_{DD} - v_{\text{out}})[(V_{GG} - V_{thl}) - \Delta V_{th}] - (V_{DD^2} - v_{\text{out}}^2)}$$

(valid for region II, that is, $|v_{\text{out}}| \geq |v_{\text{in}} - V_{thD}|$), and

$$v_{\text{in}} - V_{thD} = \frac{1}{2}v_{\text{out}} + \frac{\left\{ \begin{array}{c} -(V_{DD^2} - v_{\text{out}}^2) + (V_{DD} - v_{\text{out}})[(V_{GG} - V_{thl}) \\ + K_1(\sqrt{-(2\phi_F + v_{\text{out}})} - \sqrt{-2\phi_F}] \end{array} \right\}}{\beta_R v_{\text{out}}}$$

or $\qquad\qquad\qquad\qquad\qquad\qquad\qquad\qquad\qquad\qquad (5\text{-}16)$

$$v_{\text{in}} - V_{thD} = \frac{1}{2}v_{\text{out}} + \frac{-\frac{1}{2}(V_{DD^2} - v_{\text{out}}^2) + (V_{DD} - v_{\text{out}})[(V_{GG} - V_{thl}) - \Delta V_{th}]}{\beta_R v_{\text{out}}}$$

(valid for region III, that is, $|v_{\text{out}}| < |v_{\text{in}} - V_{thD}|$).

Again, these equations are written in terms of v_{in} as a function of v_{out} instead of vice versa, simply for convenience.

5-3   BASIC BUILDING BLOCKS

Complex MOS circuits are built from basic logic blocks such as NAND and NOR gates. Each type of gate has its own advantages and disadvantages. The peculiarities of implementing circuit logic with the MOS device is the topic of consideration in this section.

1. Simple NOR Circuit (Negative Logic). The parallel arrangement of MOS devices as seen in Fig. 5-15a lends itself well to MOS integrated circuitry. From the scale drawing of the layout (Fig. 5-15b), it is obvious that maximum utilization of space is achieved through the use of shared diffused regions and the inherent self-isolation properties of MOS devices. Notice that the three driver devices all have the same W/L ratio, making them equal g_m devices. The g_m of device Q_A is designed such that I_D/g_m yields the desired V_{ON}, as seen in the simple inverter of Sec. 5-2. Additional devices paralleling Q_A will not change the design g_m. The additional devices only tend to lower V_{ON} when more than one driver is turned on by the input logic. Figure 5-15c shows the truth table together with the basic logic equations for the NOR gate of Fig. 5-15a. (The NOR gate is so named because it is a "not-or" function.)

2. Simple NAND Circuit (Negative Logic). A series arrangement of MOS devices (Fig. 5-16a) results in a NAND logic gate. One possible layout of the three-input NAND gate is shown in Fig. 5-16b; the logic equations and truth table are given in Fig. 5-16c.

In a NAND logic gate (such as the one shown), the "not" output (complement of Q, or $\bar{Q}$) occurs when A and B and C are 1. This is equivalent to calling it a "not-and" circuit.

The ON voltage across the combination of Q_A, Q_B, and Q_C is the product of the ON current and the total resistance to ground. Since the devices are in series, their resistances add. Thus r_{ON} (total) $= 3/g_m$ for equal-geometry devices. For a given V_{ON}, the g_m (and thus the geometry) of the individual devices must be at least three times larger than for the case of a single device. As the number of devices in series increases, so do the g_m and the area. For four devices in series, each g_m must be four times greater than a single device.

3. Example.   An example will illustrate the difference in the required g_m for devices used in NAND and NOR circuits. Figure 5-17 shows a three-input combination NAND-NOR gate together with its equivalent circuit. Assuming

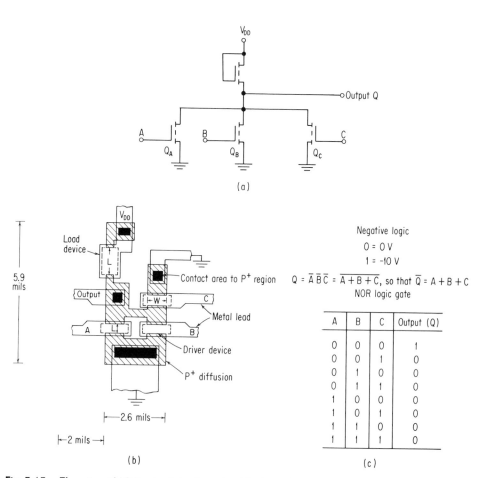

Negative logic

$0 = 0\text{ V}$
$1 = -10\text{ V}$

$Q = \bar{A}\,\bar{B}\,\bar{C} = \overline{A + B + C}$, so that $\bar{Q} = A + B + C$
NOR logic gate

A	B	C	Output (Q)
0	0	0	1
0	0	1	0
0	1	0	0
0	1	1	0
1	0	0	0
1	0	1	0
1	1	0	0
1	1	1	0

(a)

(b)

(c)

Fig. 5-15.   Three-input NOR gate: (a) schematic; (b) physical layout (scale drawing); (c) truth table.

that $V_{DD} = -15$ V, that $V_{th} = -4$ V, and that it is desired to let $|V_{ON}| \leq 0.5$ V, what must be the gain of the driver devices when $g_{ml} = 10$ μmhos? (Assume that the driver devices are driven by similar inverter stages.) From the equivalent circuit, it is seen that $|V_{ON}|$ must be ≤ 0.5 V when either A and B or C is turned on. When C is on,

$$V_{ON} \cong (V_{DD} - V_{th}) \frac{1}{2} \frac{g_{ml}}{g_{mC}}$$

[See Eq. (5-5).]

Plugging in numerical values yields

$$-0.5 \text{ V} = (-15 + 4) \frac{1}{2} \frac{10 \text{ } \mu\text{mhos}}{g_{mC}}$$

so that

$$g_{mC} = 110 \text{ } \mu\text{mhos} \tag{5-17}$$

Since V_{ON} must be the same regardless of which side is conducting, the ON voltage of the left side is equated to that of the right side, and the equation is solved for g_{mA} and g_{mB}:

$$V_{ON} \cong (V_{DD} - V_{th}) \frac{g_{ml}}{g_{mAB}} = (V_{DD} - V_{th}) \frac{1}{2} \frac{g_{ml}}{g_{mC}} \tag{5-18}$$

$$g_{mAB} = 2g_{mC} \tag{5-19}$$

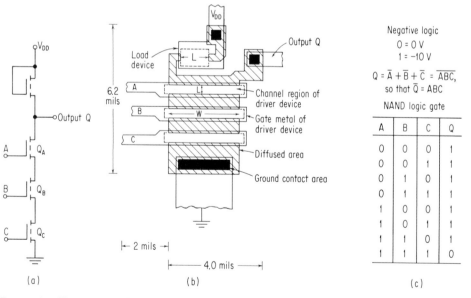

Fig. 5-16. Three-input NAND gate: (a) schematic; (b) physical layout (scale drawing); (c) truth table.

Equation (5-19) says that the g_m of devices A and B must be twice the value of device C, so that

$$g_{mA} = g_{mB} = 2 \times 110 \ \mu\text{mhos} = 220 \ \mu\text{mhos} \qquad (5\text{-}20)$$

Because of the higher g_m required, NAND-type circuitry will, in general, occupy more area than an equivalent NOR gate. This means that in the majority of cases, logic implementation by MOS technology is best done through NOR logic for the most effective utilization of area.

Variations on the theme of Fig. 5-16a can be seen in Fig. 5-18a, where three devices are used to form a four-input logic. Here the feature of a "clocked" MOS active load is shown to an advantage, where the input to the load gate is used as an input signal line. Also, the relatively high impedance levels of the MOS (as compared with the bipolar impedance levels) enables one to make use

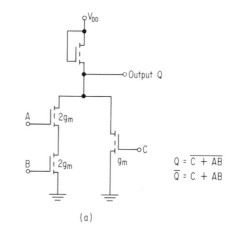

(a)

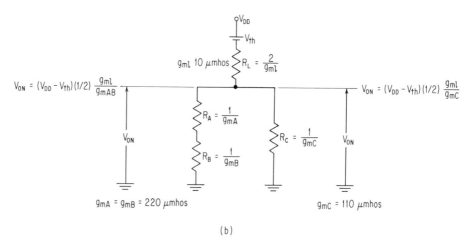

(b)

Fig. 5-17. NAND-NOR combination gate: (a) schematic; (b) equivalent ON circuit.

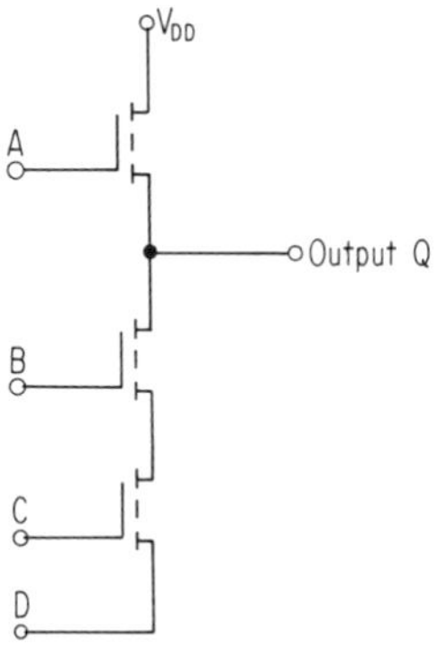

Negative logic
0 = 0 V
1 = −15 V

$$Q = A\bar{B} + AD + AB\bar{C}$$

A	B	C	D	Q
0	0	0	0	0
0	0	0	1	0
0	0	1	0	0
0	0	1	1	0
0	1	0	0	0
0	1	0	1	0
0	1	1	0	0
0	1	1	1	0
1	0	0	0	1
1	0	0	1	1
1	0	1	0	1
1	0	1	1	1
1	1	0	0	1
1	1	0	1	1
1	1	1	0	0*
1	1	1	1	1

*Because of the clocked load, this is the only logic state that draws power

(a)　　　　　　　　　　　　　(b)

Fig. 5-18. Four-input gate: (a) **schematic;** (b) **truth table.**

of the source terminal of the bottom device as an input signal line.   The logic truth table for the four-input gate is shown in Fig. 5-18b.

5-4 REDUNDANCY—IDEAL FOR MOS CIRCUITS

A unique feature of MOS technology is the ease with which redundancy can be introduced into a circuit design.*   A *series-parallel four-group* arrangement of MOS devices that is redundant on both the input and the output is shown in Fig. 5-19 for a portion of a simple inverter.   The source and drain of a given device may open or short together, and/or the gate of a given device may open or short, and the circuit will remain operative.   Whether the two center nodes connect or not is determined by the most prevalent failure mode of the device. To date, there has been no significant practical use made of redundancy.

5-5 CAPACITOR STORAGE FOR SIMPLIFIED CIRCUITRY

Another unique feature of the MOS transistor is its almost infinite input resistance.   A class of circuits which might be categorized as "dynamic logic"

* For an additional discussion of field-effect redundancy, see Ref. 2 in the Bibliography at the end of the chapter.

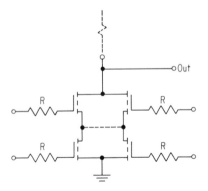

Fig. 5-19.   Redundant arrangement of MOS devices.[2,*]

can be developed by utilizing the high input resistance of an MOS in conjunction with charge stored on a capacitor.   The concept† is very simple and is illustrated in Fig. 5-20a.   An input signal is fed onto the gate of Q_1 through switch S_1. Capacitor C charges to the input voltage.   When S_1 is opened, the charge on C remains, thus biasing Q_1 either ON or OFF, depending upon the circuit conditions.   Since negligible charge is lost through the insulating gate of Q_1, the capacitor C maintains stored information—or "remembers"—until switch S_1 allows new information to be fed in.   The output from this "memory cell" is taken from the drain of Q_1, so capacitor C is never loaded down.

In actual practice, storage on C is not allowed for an infinite time.   Switch S_1 will be an MOS transistor and will exhibit some finite resistance, shunting C. The gate of Q_1 is not a perfect insulator and thus shunts another resistor across C. These two resistances (that due to S_1 dominates), together with the value of C, determine a maximum storage time for this type of circuit—frequency of operation will not extend down to direct current.   Typical lower frequency limits are in the range of 5 to 10 kHz.

An SR, utilizing the capacitor-storage concept, is shown in Fig. 5-20b.‡   The storage capacitors are shown in dashed lines.   This indicates that no particular structure is added to the IC for the storage capacitor C.   Rather, the stray capacitance of the circuit is employed as the capacitive storage element.   The system shown in the figure is a two-phase system requiring external clocks (see Fig. 5-20c).

Referring to Fig. 5-20b, assume that a logical 0 is fed into the input.   During clock time Φ_1, Q_2 and Q_3 turn on.   Because of the 0 input, P_1 is pulled up near V_{DD} by the load Q_2.   This high-voltage level, logical 1, is transferred through Q_3 onto C_1.   Clock phase Φ_2 is now turned on.   Because of the logical 1 stored on C_1, Q_4 pulls point P_3 to ground.   This information is transferred through Q_6

---

* Superscript numbers indicate works listed in the Bibliography at the end of the chapter.

† Developed by GMe/Philco and used in their SRs.

‡ From GMe/Philco Data Sheet pL5100, July, 1965.

onto C_2, where it is stored when Φ_2 goes off.   Thus a logical 0 has been shifted
from the input of the first stage to the input of the second stage.   Repeating the
above process again will shift the 0 another stage, and so on until the 0 reaches
the end of the register.   Assuming a 20-stage SR, one speaks of the 0 as having
been delayed by 20-bit lines.   A logical 1 is shifted in the same manner.

Characteristically, dynamic logic can perform a variety of given functions with
less component count than can conventional circuitry.   This reduced device

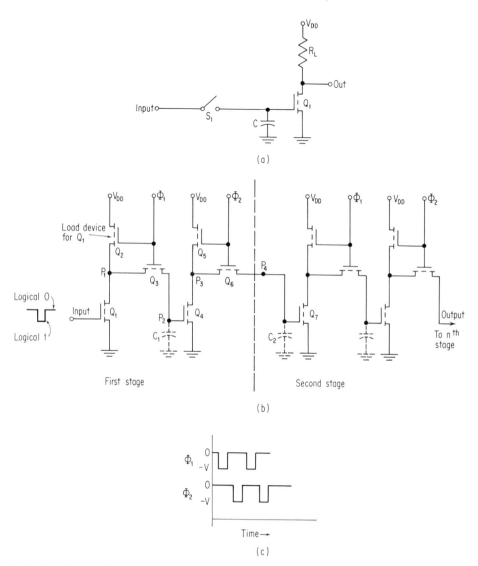

Fig. 5-20. Dynamic SR-capacitor storage: (a) basic concept; (b) schematic; (c) timing diagram of the
two clock phases.

usage results in either a given function occupying a smaller area or more functions in the same area. Ultimately, this space saving results in reduced function cost to the user. All that glitters is not gold, however, for there are drawbacks to this type of circuitry. As mentioned earlier, dynamic logic is not capable of static storage at zero frequency. Stored data must continually be circulated and shifted so as not to lose the information. A two-phase clock system must be supplied to drive the logic. At higher frequencies (approaching 1 MHz), the clock requirements may offset a good deal of the advantage gained through the use of capacitor-storage technology.

A system need not be committed entirely to the capacitor-storage technique to make use of its advantages. A very worthwhile basic SR bit can be designed by utilizing the d-c storage properties of a cross-coupled flip-flop and the sim-

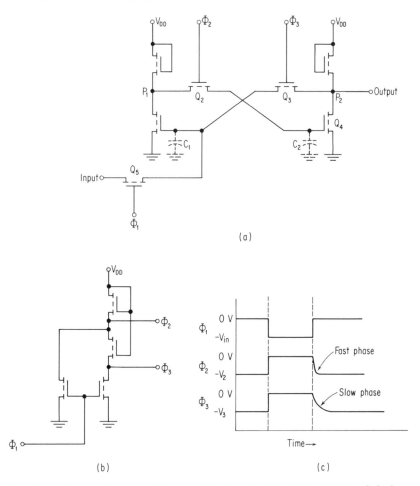

(a)

(b)   (c)

Fig. 5-21. Static SR using dynamic techniques: (a) schematic of basic bit; (b) internal clock generator; (c) timing diagram of the three clock phases.

plicity of capacitor storage for temporary memory. An SR circuit combining these two properties is shown in Fig. 5-21a.* A brief description of the circuit operation follows.

During quiescent d-c storage Φ_2 and Φ_3 hold Q_2 and Q_3 on so the circuit functions as a normal cross-coupled flip-flop. Φ_1 is at 0 V, thus holding Q_5 off and isolating the memory bit from the rest of the circuit.

During information transfer, Φ_1 turns Q_5 on while Φ_2 and Φ_3 turn Q_2 and Q_3 off—thereby isolating the right and left halves of the flip-flop. Capacitor C_2 retains its charge, so the output information of the SR bit is maintained during the clock cycle. The new input signal from the previous bit is transferred onto the gate of Q_1 (and thus C_1) through Q_5. This signal sets the output of the left inverter (P_1). At this point, Φ_1 turns off rapidly, while Φ_2 and Φ_3 begin to turn their devices on. Φ_2 is designed to be a faster clock than Φ_3, so that information in the SR will always shift forward (or to the right in Fig. 5-21a). Since clock phases Φ_2 and Φ_3 are generated on the SR chip, only Φ_1 need be supplied externally. Figure 5-21 also shows the internal clock generator and the three clock phases.

By using a three-phase clocking system, the SR bit just described is able to function as a cross-coupled flip-flop and store information down to zero frequency. Circuit complexity is reduced through the use of temporary storage during the clocking cycle. A drawback of this system is that three clock lines must run to each bit, thus presenting a topological layout problem to the IC design engineer.

5-6 MASTER–SLAVE FLIP–FLOP

The preceding section described two different SR bits utilizing capacitor storage for temporary memory. Multiple-phase clocking is used to isolate the input of the memory bit from its output in order to prevent a race or oscillating condition. This section describes a basic memory-storage cell that uses the "master-slave" concept. Here information is fed into the master flip-flop (during the clock phase) for temporary storage; information is then shifted into the slave flip-flop on the complement clock for permanent or d-c storage. Figure 5-22 outlines the step-by-step evolution of a J-K flip-flop using the master-slave technique.

Transistors Q_1 to Q_4 form a basic cross-coupled d-c flip-flop designated as the *master*. Units Q_5 to Q_8 constitute the *slave*. In Fig. 5-22b, five devices have been added to the basic cells. Q_9, Q_{10}, Q_{12}, and Q_{13} convert the basic cells to R-S flip-flops, which are interconnected so that the master controls the slave. To allow the circuit to be set to a reference position when necessary, a preset device, Q_{11}, has been added to the design. By the addition of Q_{14} and Q_{15} in Fig. 5-22c, the master and slave portions of the circuit are isolated so that a signal-race condition cannot occur. Information can now only be stored during the clock time (CP). Transfer of the information to the slave occurs only during the complement clock time ($\overline{\text{CP}}$). With the addition of the last two devices, Q_{16}

* From General Instrument Data Sheet MEM-501, May, 1965.

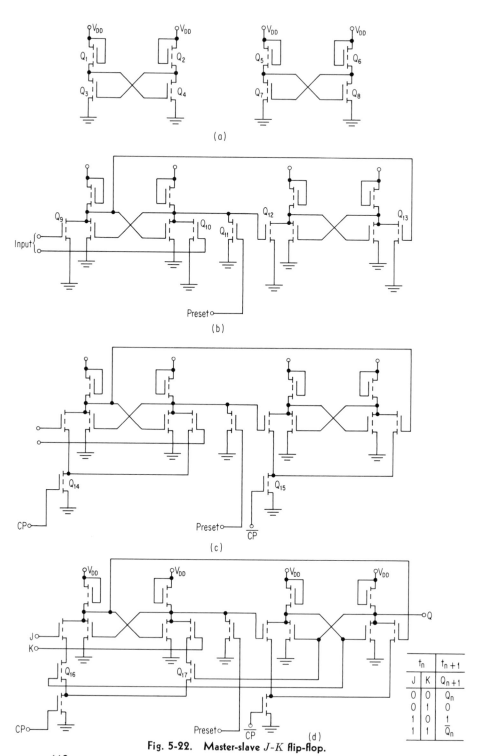

Fig. 5-22. Master-slave J-K flip-flop.

t_n		t_{n+1}
J	K	Q_{n+1}
0	0	Q_n
0	1	0
1	0	1
1	1	$\overline{Q}_n$

and Q_{17}, which convert the previous R-S flip-flop into a J-K flip-flop, the circuit is complete and is as shown in Fig. 5-22d.   A truth table is added to show the logical function of the circuit.

5-7 DESCRIPTION OF AN ACTUAL INTEGRATED CIRCUIT

MOS technology is ideally suited for the implementation of decoding matrices. A matrix array is formed by an orthogonal arrangement of diffused P+ buses (for sources, drains, grounds, and power lines) and aluminum stripes evaporated on the oxide surface (forming device gates, interconnects, and bonding pads). At each intersection where an aluminum stripe crosses over two diffused regions, there exists a possibility for an MOS transistor. Where an active device is desired, the oxide is adjusted to the appropriate thickness ($\approx 1{,}500$ Å) for a working device.   Where a transistor is undesirable in the matrix arrangement, oxide over the gate region is left thick ($\approx 10{,}000$ Å) so that the device can never turn

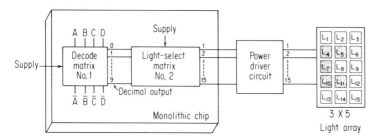

Chip size: 60 X 90 mils
Active devices: 152 transistors plus 9 protective diodes
Power: 20 - 25 mW

(a)

Code			
	A	B	C	D
---	---	---	---	---
0	0	0	1	1	$\bar{A}\bar{B}CD$
1	0	1	0	0
2	0	1	0	1
3	0	1	1	0
4	0	1	1	1
5	1	0	0	0
6	1	0	0	1
7	1	0	1	0
8	1	0	1	1
9	1	1	0	0	Excess -3 code

(b)

Fig. 5-23. Block diagram and truth table of MOS decoder.

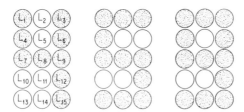

Figure 5-24

on. Thus the proper code is built into the matrix by a particular choice of active intersections.

The block diagram of a binary-to-decimal decoder illustrated in Fig. 5-23 shows that two matrices actually make up the total array. Matrix No. 1 decodes the incoming binary into a decimal output signal. The input signal consists of four inputs plus their complements, for a total of eight signal lines. Excess-3 coding was selected for matrix No. 1, whose truth table is shown in Fig. 5-23b. Matrix No. 2, the light-selection matrix, decodes the decimal output from No. 1 so that the proper combination of L_1, L_2, etc., is turned on in a 3×5 light readout array. Examples of several numerals displayed on an output matrix are shown in Fig. 5-24.

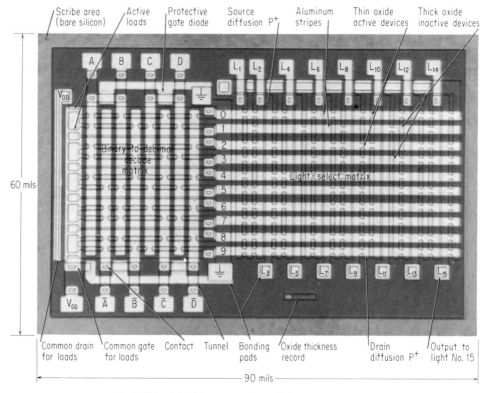

Fig. 5-25. MOS binary-to-decimal decoder complete.

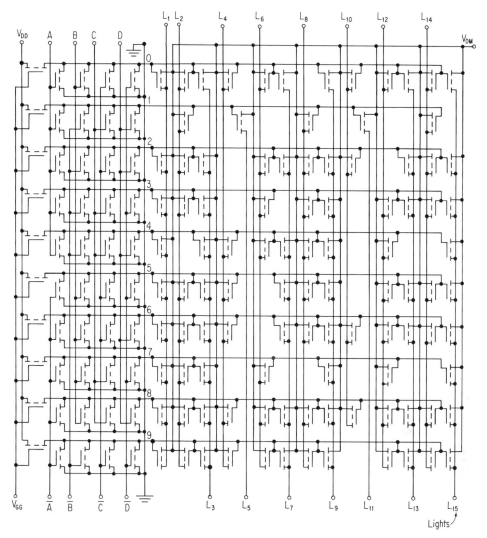

Fig. 5-26. Complete decoder schematic.

A photograph of the actual circuit is shown in Fig. 5-25. Notice how, within the matrices themselves, the aluminum leads over the oxide surface cross at right angles to the diffused regions. No crossover problems exist. When it is necessary to cross to aluminum leads (as on the decoder input), one can make use of a *tunnel*. A tunnel structure, as shown on input $\bar{D}$ (Fig. 5-25), allows one lead to "tunnel" (by way of a diffused area) under a second lead.*

A complete schematic showing all 152 active devices in the decoder array is

* For more detailed information on tunnels, or crossovers, see Ref. 3, pp. 160 and 161.

given in Fig. 5-26.   NOR-logic implementation, as shown in Fig. 5-27a, is used in the binary-to-decimal decode matrix design.   Notice that for each decimal output, only four of the possible eight devices are used (unused devices are represented in phantom lines).   The devices not used are rendered inactive by thick oxide in the gate area.   These devices are labeled in Fig. 5-25 and are shown stippled in the physical layout of the binary-to-decimal decode matrix of Fig. 5-27b.   To decode the numeral 1 requires that the 1 line be pulled "high" (-10 V) by the load device in Fig. 5-27a.   For this line to go high, the appropriate driver devices (which are controlled by the binary inputs) must be turned off.   Thus

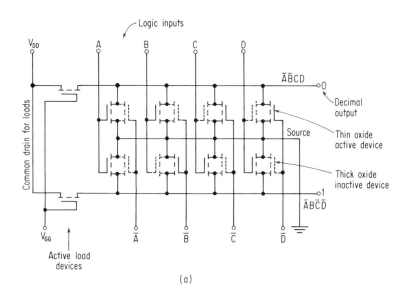

(a)

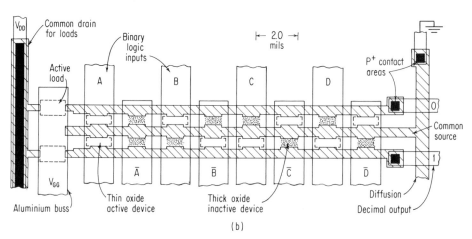

(b)

Fig. 5-27.   Binary-to-decimal decoding matrix: (a) schematic; (b) physical layout.

the $A\bar{B}CD$ lines must all be "low" (zero).   The logic equation for a 1 output as seen from both Fig. 5-27a and Fig. 5-23b is $1 = A\bar{B}CD$.

The light-selection-matrix layout is represented diagrammatically in Fig. 5-28. Inputs to this matrix are decoded decimal signals—the output from matrix No. 1. Here, as in the first matrix, thick and thin gate oxides determine the active devices. A current drive instead of a voltage swing is used for the output.   The output drains are clamped at V_{DM}, while the sources of the output devices drive the base-emitter diode of an external bipolar transistor.   In this way, the source is held near ground potential.   A current drive output such as the one just described can eliminate some of the problems normally associated with interfacing the MOS to the outside.   Certain speed advantages are also obtained from this configuration.

Considering a specific example in conjunction with Fig. 5-28 will illustrate the operation of the light-selection matrix.   Assume that a binary input signal repre-

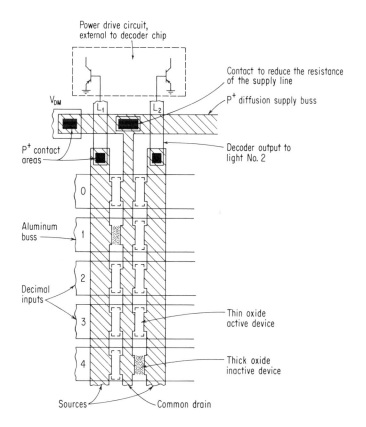

Fig. 5-28.   Light-selection matrix, physical layout.

senting the decimal 1 has been decoded by matrix No. 1.   Line 1 in the figure goes high, while the other nine lines are biased at zero.   Because of thick oxide under the 1 gate, L_1 does not turn on. L_2, however, does conduct.   Thus, in the light-output matrix (Fig. 5-24), where the center column should light for decoded 1, one can see that the MOS matrix will keep L_1 off and turn L_2 on.   Lights L_2, L_5, L_8, L_{11}, and L_{14} will light for a complete numeral.

BIBLIOGRAPHY

1. Farina, D., and D. Trotter: MOS Integrated Circuits Save Space and Money, *Electronics*, pp. 84–95, Dec. 4, 1965.
2. Wallmark, J. T., and A. G. Revesz: Redundancy in Unipolar Transistor Circuits, *IEEE Trans. Electron. Computers*, vol. EC-12, p. 23, February, 1963.
3. Engineering Staff, Motorola Inc., R. M. Warner and J. N. Fordemwalt (eds.): "Integrated Circuits," McGraw-Hill Book Company, New York, 1965.

GENERAL REFERENCES

Igarashi, R., T. Kurosawa, and T. Yarta: A 150-nanosecond Associative Memory Using Integrated MOS Transistors, *Intern. Solid-state Circuits Conf. Record*, pp. 104–105, February, 1966.
Josephs, H. C.: A Figure of Merit for Digital Systems, *Microelectronics and Reliability*, vol. 4, pp. 345–350, 1965.
Kane, J.: Switch Over to Field-Effects, Part I, *Electron. Design*, vol. 14, no. 24, pp. 54–60, Oct. 25, 1966.
———: FET's Make Digital Switching a Snap, Part II, *Electron. Design*, vol. 14, no. 25, pp. 72–79, Nov. 8, 1966.
Vadasz, L., R. Nevala, W. Sander, and R. Seeds: A Systematic Engineering Approach to Complex Arrays, *Intern. Solid-state Circuits Conf. Record*, pp. 120–121, February, 1966.

6

Analog Circuits

Field-effect devices have two distinct advantages over bipolar devices in the field of linear amplification: high input impedance and square-law transfer characteristics. High power gain (a result of the high input impedance) also makes FET's attractive amplifiers. Junction field effects are superior in noise performance when compared with the MOSFET. Small-signal linear analysis can be carried out in much the same way as pentode analysis is in sophomore electronics courses. This chapter discusses briefly a simple equivalent circuit for each of the three basic circuit configurations. Essential equations are presented along with the characteristics of each circuit configuration. The last portion of the chapter deals with various types of MOS circuits, including MOS-bipolar combinations.

6-1 COMMON–SOURCE AMPLIFIER

A simple small-signal equivalent circuit that would apply in general to MOS-FET's is shown in Fig. 6-1. (As with any device representation or equivalent circuit, the designer must decide for himself what approximations are valid at the frequency of interest. For an additional discussion of equivalent circuits, see Refs. 1 to 3 in the Bibliography at the end of the chapter.) The equivalent circuit shows that the input is essentially capacitive, that the small-signal output

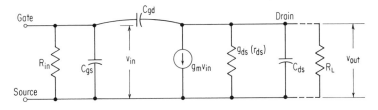

Fig. 6-1. Small-signal low-frequency equivalent circuit of an MOSFET.

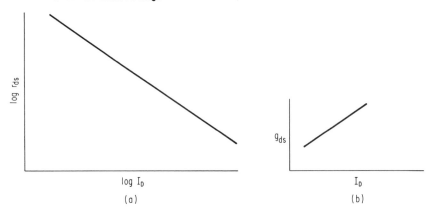

Fig. 6-2.   Output saturation immittance vs. I_D.

current is a function of the small-signal input voltage, and that the load is shunted by the output conductance of the device.   Forward gain of the device is a function of the bias point and is shown as a function of I_D in Fig. 3-2 and as a function of V_G in Fig. 3-4.   The dynamic drain resistance in the saturation region $(1/g_{ds})$ is also a function of the drain current.   The general shape of this curve is shown in Fig. 6-2a.   ($1/r_{ds}$ or g_{ds} is shown in Fig. 6-2b.)   By varying the drain-to-source distance and the substrate doping level, the curve is shifted to various locations on the graph.

Defining the voltage gain A_v as v_{out}/v_{in} in Fig. 6-1 and neglecting capacitive effects, the gain of a common-source amplifier can be written as

$$v_{out} = -g_m v_{in}(R_L \| r_{ds})$$

where the designation $R_L \| r_{ds}$ means R_L in parallel with r_{ds} and is equal to $R_L r_{ds}/(R_L + r_{ds})$.   Now

$$\frac{v_{out}}{v_{in}} = A_v = -g_m(R_L \| r_{ds})$$

or

$$A_v = \frac{-g_m}{G_L + g_{ds}} \tag{6-1}$$

Notice that when $g_{ds} \ll G_L$, the gain simplifies to

$$A_v = -g_m R_L \tag{6-2}$$

In a number of applications where a common-source stage is used, a source resistor is added for local negative feedback.   If only d-c degeneration for bias stability is desired, then the source resistor is bypassed.   However, if a-c degeneration is also desired, then a portion of the source resistance may be left unbypassed, as shown in Fig. 6-3.   Using the equivalent circuit of Fig. 6-1, to which a source resistance has been added, yields the voltage gain as a function of R_S:

$$A_v' \text{ (gain with feedback)} = \frac{A_v r_{ds}}{r_{ds} + R_L + (g_m r_{ds} + 1)R_S}$$

$$= \frac{-g_m R_L r_{ds}}{r_{ds} + R_L + (g_m r_{ds} + 1)R_S} \tag{6-3}$$

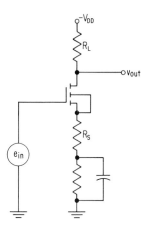

Fig. 6-3. A-c amplifier stage with source degeneration.

When g_m and r_{ds} are assumed large, the gain can be approximated by

$$A_v' = - \frac{g_m}{1 + g_m R_S} R_L \tag{6-4}$$

Notice that Eq. (6-4) is of the form $-g_m' R_L$, where g_m' is the same effective transconductance as was seen in Eq. (3-6). Thus, adding source degeneration will decrease the variation in the a-c gain by the same amount as the variation in g_m is decreased, as shown by Eq. (3-1). In Eq. (6-4), if $g_m R_S \gg 1$, then the gain further reduces to

$$A_v' \cong - \frac{R_L}{R_S} \tag{6-5}$$

This says that the gain becomes independent of device parameters. Because R_L and R_S are passive elements, they can be made very stable over wide temperature ranges, thus yielding a stable gain.

6-2 SOURCE FOLLOWER

The common-drain, or source-follower, configuration is a most useful basic circuit. Some of its properties are:

1. No phase reversal of the output signal.
2. Voltage gain always less than unity.
3. Low output impedance.
4. Large-signal swing.
5. Active impedance transformer.

Figure 6-4 shows the circuit configuration for a source follower and its equivalent circuit. Notice that the drain resistance r_{ds} is effectively in parallel with R_S. By writing this parallel combination as $r_{ds} \| R_S$, the voltage gain can be written as

$$A_v = \frac{g_m(r_{ds} \| R_S)}{1 + g_m(r_{ds} \| R_S)} \tag{6-6}$$

When $r_{ds} \gg R_S$, the parallel combination approaches R_S and the gain becomes

$$A_v = \frac{g_m}{1 + g_m R_S} R_S \qquad (6\text{-}7)$$

Equation (6-7) is in the form of $A_v = +g'_m R_S$, where g'_m is the effective transconductance of Eq. (3-6). As $g_m R_S$ becomes very large, the gain approaches unity.

Assuming a g_m of 1,000 μmhos and a source resistance of 5 kΩ, the source-follower voltage gain is +0.835. Doubling the size of R_S to 10 kΩ increases the gain to +0.91.

To find the output impedance (or conductance) of the source-follower configuration in Fig. 6-4, the gate is grounded, a signal is applied to the output, the output current is calculated, and the voltage-current ratio taken.

$$I_o = v_o(G_S + g_{ds}) + g_m v_o$$
$$\frac{I_o}{v_o} = y_o = G_S + g_{ds} + g_m \qquad (6\text{-}8)$$

where G_S = external source conductance

g_{ds} = saturated drain conductance

The output or driving impedance is thus the parallel combination of R_S, r_{ds}, and $1/g_m$. Because r_{ds} will usually be large, $r_{\text{out}} \cong R_S \| (1/g_m) = R_S/(g_m R_S + 1)$. From the previous example ($g_m = 1{,}000$ μmhos, $R_S = 5$ Ω), the output impedance $= 5(10^3)/[10^{-3}(5)(10^{-3}) + 1] = 835$ Ω. Increasing the g_m to 2,000 μmhos gives an output impedance of 445 Ω.

The source follower can be used in conjunction with a constant-current generator to obtain a gain closer to unity than is practical with reasonable resistors or power supplies. Figure 6-5 illustrates such an arrangement, which uses Q_1, R_{source}, and the Zener diode as a constant-current source driving Q_2. The incremental impedance of the current source is very high, so the a-c gain approaches unity. Yet the d-c or static resistance is low, so that excessive supplies or resistances are not necessary. Bias current is the voltage across the source resistance divided by that resistance; that is,

$$I_D = \frac{V_Z - V_{th}}{R_{\text{source}}}$$

The output impedance of this circuit approximates $1/g_{m2}$.

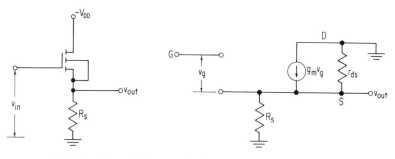

Fig. 6-4. Source follower: (a) schematic; (b) equivalent circuit.

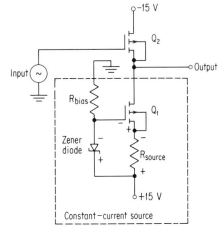

Fig. 6-5. Source-follower stage biased by a constant-current generator.

6-3 COMMON–GATE AMPLIFIER (GROUNDED GATE)

A common-gate stage is analogous to the common-base stage in the world of bipolar transistors. This configuration offers impedance transformation in the opposite direction to that of the source follower—namely, low impedance to high.

Figure 6-6 shows the grounded-gate stage together with its equivalent circuit. The input impedance as calculated from this circuit is

$$r_{\text{in}} = \frac{R_L + r_{ds}}{1 + g_m r_{ds}}$$

If $g_m r_{ds} \gg 1$,

$$r_{\text{in}} = 1/g_m + \frac{R_L}{g_m r_{ds}} \tag{6-9}$$

and is just the channel resistance $(1/g_m)$ plus a term due to the load resistor $(R_L/g_m r_{ds})$. A common-gate stage has a positive voltage gain, indicating that the input and output signals are in phase. The gain is given as

$$A_v = \frac{+(1 + r_{ds}g_m)R_L}{R_L + r_{ds} + (1 + r_{ds}g_m)R_S} \tag{6-10}$$

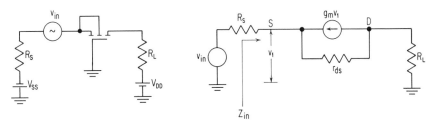

Fig. 6-6. Common-gate amplifier and equivalent circuit.

If $R_S \ll (R_L + r_{ds})/(1 + g_m r_{ds})$ (the output impedance) and $r_{ds} \gg R_L$, then the gain reduces to

$$A_v \cong +g_m R_L \tag{6-11}$$

6-4 VARIOUS AMPLIFIER CONFIGURATIONS

1. Self-biasing. In a P-channel enhancement MOS, the gate must be biased negatively to initiate conduction. This is conveniently done by returning the gate to its drain terminal through a large feedback resistor (see Fig. 6-7a). Since no d-c current is drawn, the full drain voltage is impressed on the gate. Analysis is greatly simplified when compared with the bipolar case, where an IR drop must be considered. Figure 6-7b shows the characteristic curves of the driver device, together with (1) the locus of points where $V_{DG} = V_{GS}$, and (2) the static load line. The quiescent operating point is located at the intersection of these two curves. Bias changes can be carried out effectively by varying R_L, while variations R_f have no effect.

D-c feedback in Fig. 6-7a results in stability of the stage greater than that of a single device biased to a fixed supply. If external conditions cause the drain current to decrease, the drain terminal is pulled closer to the supply—thus putting more negative voltage on the gate. The negative voltage tends to increase the current, shifting operation back toward the original bias point. A-c feedback is also achieved with the illustrated configuration. This feedback can be eliminated by splitting R_f in half and grounding the resulting node through a capacitor, or it can be enhanced by paralleling R_f with C_f.

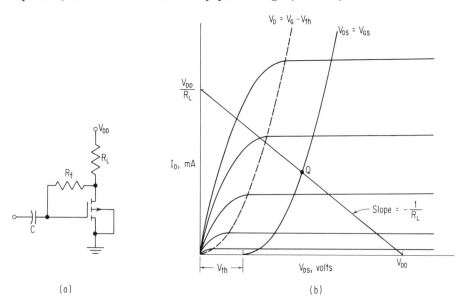

(a)   (b)

Fig. 6-7. Feedback biasing: (a) schematic; (b) graphical solution of operating point.

2. Large-signal-amplifier Stage. Section 5-2 described the operation of the inverter stage shown in Fig. 6-8a. Without duplicating the efforts presented there, the results will be presented as they apply to the linear-amplifier case.

Figure 6-8b shows the transfer curve describing inverter operation. Region II describes the area of linear operation. Here, both the load and driver devices are in the saturation region. The inverter configuration is such as to allow cancellation of their nonlinear (square-law) characteristics, resulting in a linear portion of the transfer curve. Gain in this region is dependent upon the square root of the β ratios of the two devices. All terms except W/L ratios cancel, resulting in the following expression for the a-c gain:

$$A_v = -\sqrt{\frac{(W/L)_D}{(W/L)_l}} \qquad (6\text{-}12)$$

Notice that the result is dependent only upon geometrical terms and is completely independent of such temperature-sensitive terms as device mobility. This suggests that the configuration of Fig. 6-8a may make a good temperature-stable a-c amplifier. Data taken on an inverter with an approximate voltage gain of 2 are presented in graphical form in Fig. 6-9. Gain variation is less than 2 percent over a 105°C temperature swing, which is a remarkable result when one considers, first, that semiconductors are extremely temperature-sensitive by their very nature, and second, that no external compensation or feedback was used.

To demonstrate the linear property of the MOS inverter, three different types of signals were fed into the inverter stage shown in Fig. 6-10. (This represents the IC case where the load substrate is returned to ground.) Reasonably faithful representation of the 1-kHz input waveforms is observed.

3. MOS-bipolar Combinations. Large values of transconductance can be obtained through the use of an MOS and a bipolar in a *Darlington* type of con-

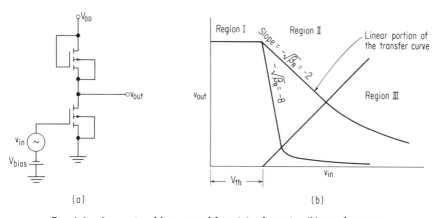

(a)   (b)

Fig. 6-8. Large-signal linear amplifier: (a) schematic; (b) transfer curve.

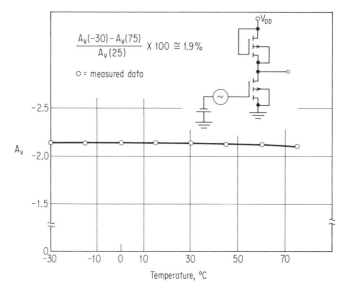

Fig. 6-9. Temperature characteristic of MOS large-signal amplifier.

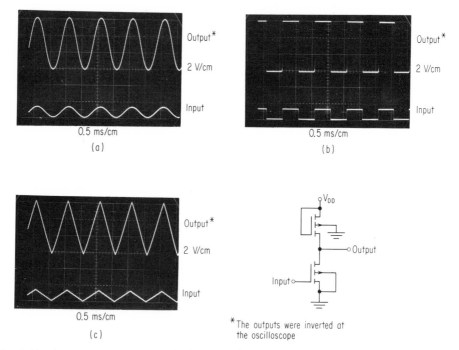

Fig. 6-10. Input-output wave shapes for large-signal amplifier: (a) sine-wave input; (b) square-wave input; (c) triangular-wave input.

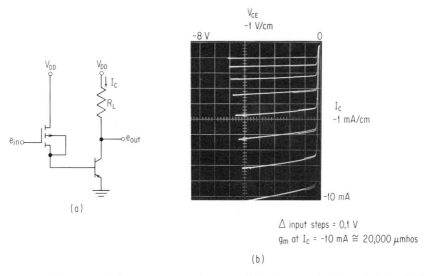

Fig. 6-11. MOS-bipolar Darlington-type configuration: (a) schematic; (b) characteristics displayed on curve tracer.

figuration.*   Gain multiplication results in transconductances reaching many tens of thousands of μmhos.   An arrangement such as that shown in Fig. 6-11a combines the most desirable properties of each device into one composite, i.e., the high input impedance of the field effect together with the low V_{sat} and high g_m of the bipolar device.   A characteristic curve as seen on a curve tracer for this configuration is shown in Fig. 6-11b.

Cascade arrangements (such as that shown in Fig. 6-12a) offer versatility in

* For an additional discussion on ways to pair FET's and bipolars, see Ref. 4 in the Bibliography at the end of the chapter.

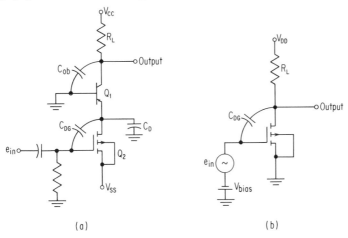

Fig. 6-12. MOS-bipolar cascade arrangement: (a) inverter stage; (b) simple MOS inverter.

design as well as improved frequency performance. Driving the MOS into the emitter of a common-base bipolar stage (a virtual short circuit) keeps the voltage swing of the drain to a minimum. With zero voltage gain, capacitance C_{DG} is not enhanced by the Miller effect, and virtually no current is lost in charging C_{DG} and C_D. Frequency response of the bipolar is high because it is biased as a common-base amplifier (here again, the Miller effect is significantly reduced). Frequency response of the cascade arrangement is determined by the $R_L C_{ob}$ time constant (assuming the intrinsic limit of the devices has not been reached).

A higher output impedance (by an order of magnitude) can be expected from the cascade configuration than from the simple MOS inverter. Channel-length modulation results in a lower dynamic drain resistance* for the case of Fig. 6-12b than the situation in Fig. 6-12a. The gain of either stage in Fig. 6-12 is $-g_m R'_L$;† however, the lower output resistance of the simple inverter stage shunts the external load, resulting in a lower a-c gain.

Figure 6-13 shows a complementary feedback pair that offers design simplicity and stable operation.[5],‡ When the gains of the two devices are large, the overall gain of the feedback pair becomes independent of the transistor parameters and can be approximated by the expression

$$A_v = +\frac{R_1 + R_2}{R_1} \tag{6-13}$$

Gains in the order of 10 to 20 can be obtained from practical circuits connected in the feedback-pair arrangement.

Unity-gain amplifiers are very useful in applications requiring impedance transformation or isolation stages. By setting $R_2 = 0$, the gain of the amplifier approaches unity, as seen from Eq. (6-13). This modified source follower has an effective source resistance of $h_{FE} \times R_1$. Equation (6-7) describes the gain of a source follower, which can be written in the following way:

$$A_v = \frac{R_S}{(1/g_m) + R_S} \tag{6-14}$$

Assuming a g_m of 1,000 μmhos for the input device, $R_1 = 1,000\ \Omega$, and an h_{fe} for the bipolar of 100 results in the gain

$$A_v = \frac{100 \times 1\ \text{k}\Omega}{1,000\ \Omega + 100 \times 1\ \text{k}\Omega} = \frac{100}{101} = 0.99$$

This is close to unity and agrees with the gain approximation presented in Eq. (6-13). Obviously, R_2 can be trimmed to produce a gain as close to unity as is required.

---

* See Sec. 6-1.

† $R'_L = R_L \| r_{\text{out}}$.

‡ Superscript numbers indicate works listed in the Bibliography at the end of the chapter.

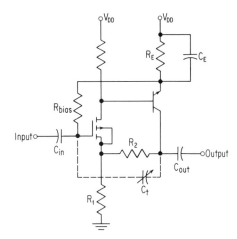

Fig. 6-13. MOS-bipolar complementary-feed-back amplifier.

The high-frequency performance of the stage shown in Fig. 6-13 can be improved by the addition of a small trimming capacitor, C_t, from the output to the input. This is a positive feedback path which cancels the shunting effect of stray capacitance at the input.

Because no current flows in the gate circuit, the input resistance of the amplifier can be made quite large by increasing R_{bias} up to 100 MΩ or larger. Bias stability is improved because of the two feedback paths, effective source resistance $R_1 h_{FE}$ and R_{bias} returned to the emitter of the bipolar.

4. Phase Splitter. Often it is desirable to split a signal into two signals 180° apart but of the same amplitude. Field-effect transistors are ideally suited for this—more so than are bipolar devices. The basic amplifier is shown in Fig. 6-14. Because the same current flows through R_D and R_S, the gains at the two output nodes must be equal if the resistors are equal. In the bipolar case, how-

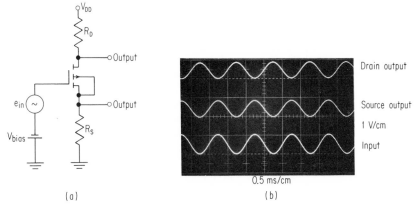

(a)   (b)

Fig. 6-14. MOS phase splitter: (a) circuit; (b) input-output waveforms.

ever, the collector and emitter currents are not exactly equal. This produces a situation of unequal gains when equal resistors are used.

Equation (6-7) specifies the gain at the source terminal as

$$\text{Gain}_{\text{source}} = + \frac{g_m}{1 + g_m R_s} R_S \tag{6-15}$$

Equation (6-4) specifies the gain at the drain terminal as

$$\text{Gain}_{\text{drain}} = - \frac{g_m}{1 + g_m R_S} R_D \tag{6-16}$$

These equations can be written in the form $g'_m R$, so that when $R_S = R_D = R_L$, the gain expression for each output node can be given as

$$A_v = \pm g'_m R_L \tag{6-17}$$

Figure 4-14 shows a scope photo of the outputs from a phase splitter with equal resistors and of the input signal.

BIBLIOGRAPHY

1. Wallmark, J. T.: The Field-effect Transistor—A Review, *RCA Rev.*, pp. 641–660, December, 1963.
2. Fischer, W.: Equivalent Circuit and Gain of MOS Field Effect Transistors, *Solid-state Electron.*, vol. 9, pp. 71–81, 1966.
3. Lukes, Z.: Characteristics of the Metal-oxide-semiconductor Transistor in the Common-gate Electrode Arrangement, *Solid-state Electron.*, vol. 9, pp. 21–27, 1966.
4. Parmer, W. F.: Four Ways to Pair Field-effect and Conventional Transistors, *Electron. Design*, pp. 44–47, Aug. 16, 1962.
5. Crawford, B.: Complementary Two-stage Amplifiers, *Electro-technol.*, pp. 47–53, May, 1964.

GENERAL REFERENCES

Blaser, L.: Dual MOS-FET Simplifies FM Multiplex Decoder, *Electron. Design*, pp. 78–79, March 1, 1966.

Freshour, S. G.: Capacitively Tuned FET and MOS Oscillators, *solid/state/design*, pp. 28–32, December, 1965.

Luettgenau, G. G., and S. H. Barns: Designing with Low-noise MOSFET's: A Little Different but No Harder, *Electronics*, pp. 53–57, Dec. 14, 1964.

Phalan, J. M.: MOSFET's Give Long Time-constant Ramps, *EEE*, p. 46, April, 1966.

Seashore, C. R.: FET Audio Signal Mixer Exhibits Linearity, Isolation, *Electron. Design*, p. 242, March 15, 1966.

Skopal, T.: MOS-FET Circuit Stores Input Voltage Peaks as D.C., *Electron. Design*, p. 76, Feb. 1, 1966.

White, M. H.: A Voltage-controlled MOS-FET Integrator, *Proc. IEEE, Correspondence*, pp. 421–422, March, 1966.

Appendix

Conversion Table

$$1 \text{ centimeter (cm)} = 10^8 \text{ Å}$$
$$= 10^4 \text{ }\mu$$
$$= 4 \times 10^2 \text{ mils}$$

$$1 \text{ angstrom (Å)} = 3.937 \times 10^{-6} \text{ mil} \cong 4 \times 10^{-6} \text{ mil}$$
$$= 1 \times 10^{-8} \text{ cm}$$
$$= 1 \times 10^{-4} \text{ }\mu$$

$$1 \text{ micron }(\mu) = 3.937 \times 10^{-2} \text{ mil} \cong 4 \times 10^{-2} \text{ mil}$$
$$= 1 \times 10^{-4} \text{ cm}$$
$$= 1 \times 10^4 \text{ Å}$$

$$1 \text{ Millinch mil} = 1 \times 10^{-3} \text{ in}$$
$$= 2.5 \times 10^{-3} \text{ cm}$$
$$= 2.54 \times 10^5 \text{ Å} \cong 2.5 \times 10^5 \text{ Å}$$
$$= 2.54 \times 10 \text{ }\mu \cong 25 \text{ }\mu$$

Gate
$$900 \text{ Å} \cong 3.6 \times 10^{-3} \text{ mil}$$
$$1{,}000 \text{ Å} \cong 4 \times 10^{-3} \text{ mil}$$
$$1{,}500 \text{ Å} \cong 6 \times 10^{-3} \text{ mil}$$

Channel length
$$5 \text{ }\mu \cong 2 \times 10^{-1} \text{ mil}$$
$$10 \text{ }\mu \cong 4 \times 10^{-1} \text{ mil}$$
$$12.5 \text{ }\mu \cong 5 \times 10^{-1} \text{ mil}$$

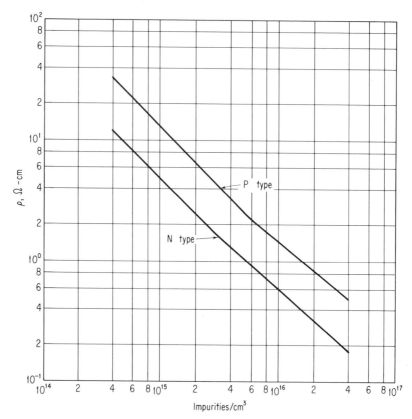

Fig. A-1. Resistivity vs. impurity density in silicon. *(From J. C. Irvin, Resistivity of Bulk Silicon and of Diffused Layers in Silicon, Bell System Tech. J., vol. 41, pp. 387–410, 1962. By permission of the American Telephone and Telegraph Company.)*

Index

Index